Concise Introduction to Logic and Set Theory

Concise Introduction to Logic and Set Theory

Iqbal H. Jebril, Hemen Dutta, and Ilwoo Cho

CRC Press
Taylor & Francis Group
Boca Raton London New York

CRC Press is an imprint of the
Taylor & Francis Group, an **informa** business

First edition published 2022
by CRC Press
6000 Broken Sound Parkway NW, Suite 300, Boca Raton, FL 33487-2742

and by CRC Press
2 Park Square, Milton Park, Abingdon, Oxon, OX14 4RN

Library of Congress Cataloging-in-Publication Data

Names: Jebril, Iqbal H., author. | Dutta, Hemen, 1981- author. | Cho, Ilwoo, author.
Title: Concise introduction to logic and set theory / Iqbal H. Jebril, Hemen Dutta, and Ilwoo Cho.
Description: First edition. | Boca Raton : CRC Press, 2022. | Series: Mathematics and its applications: modelling, engineering, and social sciences | Includes bibliographical references and index.
Identifiers: LCCN 2021020846 (print) | LCCN 2021020847 (ebook) | ISBN 9780367077952 (hardback) | ISBN 9781032106229 (paperback) | ISBN 9780429022838 (ebook)
Subjects: LCSH: Logic, Symbolic and mathematical. | Set theory.
Classification: LCC QA9.A5 J43 2022 (print) | LCC QA9.A5 (ebook) | DDC 511.3--dc23
LC record available at https://lccn.loc.gov/2021020846
LC ebook record available at https://lccn.loc.gov/2021020847

ISBN: 978-0-367-07795-2 (hbk)
ISBN: 978-1-032-10622-9 (pbk)
ISBN: 978-0-429-02283-8 (ebk)

DOI: 10.1201/9780429022838

Typeset in Nimbus Roman
by KnowledgeWorks Global Ltd.

Contents

Preface

The book deals with two most important branches of mathematics, namely, logic and set theory. Logic and set theory are two closely related branches of mathematics that play very crucial role in the foundations of mathematics, and together these produced several beautiful results in all of mathematics. The book is designed for various courses where mathematical logic and set theory are required either as compulsory subjects or as parts of other subjects. The book consists of five chapters, and they are organised as follows:

Chapter 1 deals with basics of mathematical logic. The chapter starts with the concept of "set" briefly with examples. Then several topics and notions, such as propositions, connectives, tautology, contradiction, quantifiers, logical reasoning, mathematical proof, direct and contrapositive proof, contradiction proof, and induction principle, are discussed with examples. Exercises are also incorporated at different places of the chapter.

Chapter 2 starts with the basic notions of set theory. Then several elementary properties and operations on sets, such as union of sets, intersection of sets, complement of set, and difference and symmetric difference of sets, are discussed with examples and diagrammatic representations. Finally, the chapter discusses indexed families of sets. Several exercises relevant to the topics discussed are also incorporated at different places within the chapter.

Chapter 3 deals with the notion of relations with lots of examples, diagrammatic representations, and exercises. In particular, the chapter discusses notions such as ordered pairs and Cartesian product, relations on sets, type of relations, equivalence relations, equivalence classes, congruence, and partial and total ordered relations.

Chapter 4 discusses the concept of functions, and basic ideas associated with this notion with the help of several examples and diagrammatic representations. In particular, the chapter discusses concepts such as domain and range of a function, graph of a function, functions that are onto and one-to-one, composite function, inverse function, and direct images and inverse images under function. Relevant exercises are also added at different locations of the chapter.

Chapter 5 deals with the cardinality of sets. Several notions and associated ideas on finite and infinite sets, equivalent sets, countable and uncountable sets, and cardinal arithmetic are also discussed. Finally, Cantor's theorem, Schröder–Bernstein theorem and axiom of choice are presented. This chapter also contains several examples and exercises.

We would like to offer our sincere thanks to all the authors who have contributed extensively in the field of set theory and mathematical logic and our family members and friends who have encouraged us to develop this book. The authors are also grateful to Ants Aasma (Estonia) for carefully reading/editing the book and making several fruitful suggestions to improve the presentation of the book.

Iqbal H. Jebril
Hemen Dutta
Ilwoo Cho
March 2021

About Authors

Iqbal H. Jebril is Professor in the Department of Mathematics at Al-Zaytoonah University of Jordan, Amman, Jordan. He obtained his Ph.D. from the National University of Malaysia (UKM), Malaysia. His fields of research interest include functional analysis, operator theory, and fuzzy logic. He has several prestigious journal and conference publications to his credit. He is also serving for several journals and conferences in different capacities.

Hemen Dutta obtained his M.Phil and Ph.D. both in mathematics and also completed postgraduate Diploma in Computer Application from Gauhati University, India. He is a regular teaching faculty member in the Department of Mathematics at Gauhati University, India. His current research interests include topics in nonlinear analysis and mathematical modelling. He is currently a regular and guest editor of several SCI/SCIE indexed journals. He has also published several thematic issues in leading journals and books with reputed publishers.

Ilwoo Cho obtained his Ph.D. from the Department of Mathematics, University of Iowa, USA. His major fields of research are free probability, operator theory, operator algebra, and dynamical systems. He is full professor in the Department of Mathematics & Statistics at Saint Ambrose University, USA. He is also an assistant editor of *Complex Analysis & Operator Theory* and Mathematics Reviewer of the American Mathematical Society.

1 Mathematical Logic

Logic is the analysis of reasoning, and it concerns with the picture of thought not with its subject. Mathematical logic, however, is one of the mathematical branches that concerns with explaining types of reasoning used by mathematicians.

In this chapter, we present set concept briefly. Set theory is the cornerstone in the foundations of mathematical logic, and it is the device of defining and analyzing primary concepts. After that we define the proposition and explain the way to connect simple propositions with one link or more to get a compound proposition. Then, we talk about the qualified propositions and valid arguments. We conclude the chapter with methods of mathematical proof.

1.1 SETS

In this section, we present a simple introduction of sets, which form a base of studying logic. In chapter two, we will describe set theory in more detail, giving notations and results, which we need later.

Although we will not formally define the word "set", we will use it to refer to a collection of objects of some sort. We indicate that an object x is in a set A by writing $x \in A$. If x is not in A, we write $x \notin A$. The objects in a set A are usually called the **elements** of A.

Example 1.1.1

(1) $2 \in \{1,2,3,a,b\}$,

(2) $3 \notin \{-1,0,1\}$.

We say that sets A and B are equal and write $A = B$ in case A and B contain exactly the same elements.

Example 1.1.2
$$\left\{0, \frac{\sqrt{3}}{2}, 6, 1\right\} = \left\{1, 6, \sin 2\pi, \cos \frac{\pi}{6}\right\}.$$

We shall use the following notations for some common sets of numbers,

$\mathbb{N} = \{1,2,3,\dots\}$, the set of **natural numbers.**

$\mathbb{Z} = \{0,\pm 1,\pm 2,\pm 3,\dots\}$, the set of **integers.**

$\mathbb{Q} = \{\frac{a}{b} : a,b \in \mathbb{Z}, b \neq 0\}$, the set of **rational numbers.**

DOI: 10.1201/9780429022838-1

$\mathbb{R} =$ the set of **real numbers.**

$\mathbb{C} = \{a + bi : a, b \in \mathbb{R}, i^2 = -1\}$, the set of **complex numbers.**

$\mathbb{Z}^+ = \{1, 2, 3, \ldots\}$, the set of **positive integers.**

$\mathbb{Z}^- = \{-1, -2, -3, \ldots\}$, the set of **negative integers.**

$\mathbb{Z}_e = \{0, \pm 2, \pm 4, \ldots\}$, the set of **even integers.**

$\mathbb{Z}_o = \{\pm 1, \pm 3, \pm 5, \ldots\}$, the set of **odd integers.**

Let $a, b \in \mathbb{R}$ such that $a < b$. Then,
the open interval (a, b) is the set of real numbers between a and b, but does not include the endpoints a and b, that is

$$(a, b) = \{x \in \mathbb{R} \mid a < x < b\},$$

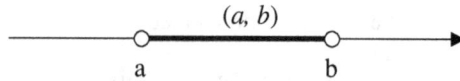
(a, b)

Whereas, the closed interval $[a, b]$ is the set of real numbers between a and b, and it includes the endpoints a and b, that is

$$[a, b] = \{x \in \mathbb{R} \mid a \leq x \leq b\},$$

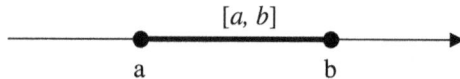
$[a, b]$

the half-open interval (or half closed) $[a, b)$ & $(a, b]$

$$[a, b) = \{x \in \mathbb{R} \mid a \leq x < b\}, x = a, x \neq b.$$

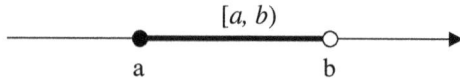
$[a, b)$

$$(a, b] = \{x \in \mathbb{R} \mid a < x \leq b\},$$

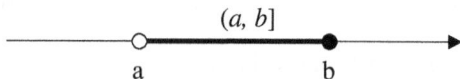
$(a, b]$

Note that

$$[a,\infty) = \{x \in \mathbb{R} \mid x \geq a\}, (a,\infty) = \{x \in \mathbb{R} \mid x > a\},$$

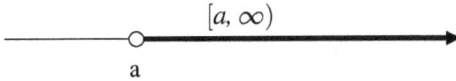

$$[a, \infty)$$

a

$$(-\infty,a] = \{x \in \mathbb{R} \mid x \leq a\}, (-\infty,a) = \{x \in \mathbb{R} \mid x < a\}.$$

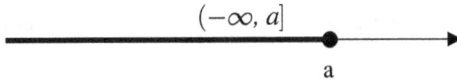

$$(-\infty, a]$$

a

1.2 PROPOSITIONS AND CONNECTIVES

In English language, some sentences are interrogatory (Where is my book?), others exclamatory (Oh!), and others have a definite sense of truth to them.

Definition 1.2.1 A proposition (or statement) is a sentence that is either true or false. Thus, a proposition has exactly one truth value: true, which we denote by T, or false, which we denote by F.

Some examples of propositions are

(a) $3 + 4 = 7$ (true proposition),

(b) $\sqrt{3}$ is a rational number (false proposition),

(c) Amman is the capital of Jordan (true proposition),

(d) If $f(x) = \sin x$, then $f'(x) = \cos x$ (true proposition),

(e) There exists a natural number n such that $n < 0$ (false proposition), and

(f) Al Riyadh is the capital of Saudi Arabia (true proposition).

The following sentences are not propositions:

(g) How are you?

(h) $x + 9 = 100$.

(i) Give me your book.

Remark 1.2.2 Propositions will be denoted by capital letters like:

$$P, Q, R, S, \ldots$$

Definition 1.2.3 Any two propositions can be combined by the word "and" to form a composite proposition, which is called conjunction of the original propositions. Symbolically, the conjunction of two propositions P and Q is denoted by $P \wedge Q$.

Example 1.2.4 Let P be "it is raining" and let Q be "the sun is shining". Then $P \wedge Q$ denotes the proposition "it is raining and the sun is shining".

Example 1.2.5 Let P be "$3 + 2 \neq 7$", and let Q be "Paris is in France". Then $P \wedge Q$ denotes the proposition "$3 + 2 \neq 7$ and Paris is in France".

The truth value of the composite proposition $P \wedge Q$ satisfies the following property:

Property 1.2.6 If P is true and Q is true, then $P \wedge Q$ is true; otherwise $P \wedge Q$ is false. In other words, the conjunction $P \wedge Q$ of two propositions P and Q is true if and only if both components P and Q are true.

Examples 1.2.7 Consider the following four propositions:

(a) Damascus is in Syria and $4 + 5 = 10$,

(b) Damascus is in Syria and $4 + 5 = 9$,

(c) Damascus is in Syria and $4 + 5 = 0$,

(d) Damascus is in Syria and $4 + 5 = -9$.

By the property 1.2.6, only (b) is true.

A convenient way to state the above property is by means of a truth table as follows:

P	Q	$P \wedge Q$
T	T	T
T	F	F
F	T	F
F	F	F

Note that the first line is a short way of saying that if P is true and Q is true, then $P \wedge Q$ is true. The other lines have analogous meaning.

Definition 1.2.8 Any two propositions can be combined by the word "or" to form a new proposition, which is called disjunction of the original two propositions. Symbolically, the disjunction of propositions P and Q is denoted by $P \vee Q$.

Any two propositions can be combined by the word "or" to form a new proposition, which is called the disjunction of the original two propositions. Symbolically, the disjunction of propositions P and Q is denoted by $P \vee Q$.

Example 1.2.9 Let P be "$1 \neq 5$" and let Q be "$\sqrt{2}$ is irrational". Then $P \vee Q$ is the proposition "$1 \neq 5$ or $\sqrt{2}$ is irrational".

The truth value of the composite proposition $P \vee Q$ satisfies the following property:

Property 1.2.10 If P is true or Q is true or both P and Q are true, then $P \vee Q$ is true; otherwise, $P \vee Q$ is false. In other words, the disjunction of two propositions is false only if each component is false.

Property 1.2.10 can be written in the form of table as follows:

P	Q	$P \vee Q$
T	T	T
T	F	T
F	T	T
F	F	F

Definition 1.2.11 Given any proposition P, another proposition, called **negation** of P, can be formed by writing "it is false that ..." before P, or, if possible, by adding the word "not" to P. Symbolically, the negation of P is denoted by $\sim P$.

Example 1.2.12 Consider the following propositions:

(a) Amman is in Jordan,

(b) It is false that Amman is in Jordan,

(c) Amman is not in Jordan.

Then (b) and (c) are each the negation of (a).

Example 1.2.13 Consider the following propositions:

(a) $3 + 4 = 9$,

(b) It is false that $3 + 4 = 9$,

(c) $3 + 4 \neq 9$.

Then both (b) and (c) are the negation of (a).

The truth value of the negation of a proposition satisfies the following property:

Property 1.2.14 If P is true, then $\sim P$ is false, and if P is false, then $\sim P$ is true. In other words, the truth value of the negation of a proposition is always the opposite of the truth value of the original proposition.

Property 1.2.14 can also be written in the form of a table as follows:

P	$\sim P$
T	F
F	T

Example 1.2.15 Consider the propositions in Example 1.2.12 above. Notice that (a) is true, and (b) and (c), its negations, are false.

Example 1.2.16 Consider the propositions in Example 1.2.13 above. Notice that (a) is false, and (b) and (c), its negations, are true.

Definition 1.2.17 Many propositions, especially in mathematics, are of the form "if P, then Q". Such propositions are called conditional propositions and are denoted by $P \Rightarrow Q$.

The conditional $P \Rightarrow Q$ can also be read:

(1) P implies Q,

(2) P only if Q,

(3) P is sufficient for Q,

(4) Q is necessary for P.

The truth value of the conditional proposition $P \Rightarrow Q$ satisfies the following property:

Property 1.2.18 The conditional $P \Rightarrow Q$ is true unless P is true and Q is false. In other words, property 1.2.18 states that a true proposition cannot imply a false proposition.

Property 1.2.18 can be written in the form of a table as follows:

P	Q	$P \Rightarrow Q$
T	T	T
T	F	F
F	T	T
F	F	T

Example 1.2.19 Consider the following propositions:

(a) If Al Riyadh is in Saudi Arabia, then $7 + 2 = 13$,

(b) If Al Riyadh is in Saudi Arabia, then $7 + 2 = 9$,

(c) If Al Riyadh is in Jordan, then $7 + 2 = 9$,

(d) If Al Riyadh is in Jordan, then $7 + 2 = 13$.

By the property 1.2.18, only (a) is a false proposition. The others are true.

Definition 1.2.20 Another composite proposition is of the form "P if and only if Q" or (shortly) "P iff Q". Such propositions are called bi-conditional propositions and are denoted by $P \Leftrightarrow Q$.

The truth value of the bi-conditional proposition $P \Leftrightarrow Q$ satisfies the following:

Property 1.2.21 If P and Q have the same truth value, then $P \Leftrightarrow Q$ is true; if P and Q have opposite truth values, then $P \Leftrightarrow Q$ is false.

Property 1.2.21 can be written in the form of a table as follows:

P	Q	$P \Leftrightarrow Q$
T	T	T
T	F	F
F	T	F
F	F	T

Example 1.2.22 Consider the following propositions:

(a) $2 < 4 \Leftrightarrow -1 < 0$,

(b) $2 + 3 = 5 \Leftrightarrow$ Amman is in Iraq,

(c) $\sqrt{2}$ is a rational number $\Leftrightarrow$ Amman is in Jordan,

(d) -3 is a natural number $\Leftrightarrow$ Baghdad is in Jordan.

According to property 1.2.21, (a) and (d) are true; however; (b) and (c) are false.

Remark 1.2.23 The following expressions give the same meaning:

(a) $P \Leftrightarrow Q$,

(b) P is necessary and sufficient for Q,

(c) Q is necessary and sufficient for P,

(d) P if and only if Q.

Definition 1.2.24 Two propositions P and Q are said to be logically equivalent if their truth tables are identical. We denote the logical equivalence of P and Q by $P \equiv Q$.

Example 1.2.25 The truth tables of $(P \Leftrightarrow Q) \wedge (Q \Rightarrow P)$ and $P \Leftrightarrow Q$ are as follows:

P	Q	$P \Rightarrow Q$	$Q \Rightarrow P$	$(P \Leftrightarrow Q) \wedge (Q \Rightarrow P)$
T	T	T	T	T
T	F	F	T	F
F	T	T	F	F
F	F	T	T	T

P	Q	$P \Leftrightarrow Q$
T	T	T
T	F	F
F	T	F
F	F	T

Hence $(P \Leftrightarrow Q) \wedge (Q \Rightarrow P) \equiv P \Leftrightarrow Q$.

Example 1.2.26 The truth tables of $P \Rightarrow Q$ and $\sim P \vee Q$ are as follows:

P	Q	$P \Rightarrow Q$
T	T	T
T	F	F
F	T	T
F	F	T

P	Q	$\sim P$	$\sim P \vee Q$
T	T	F	T
T	F	F	F
F	T	T	T
F	F	T	T

Hence $(P \Rightarrow Q) \equiv \sim P \vee Q$.

Exercises

1. Which of the following sentences are propositions? Justify your answer.

    (a) $x < 5$.

    (b) $x + y = y + x$.

    (c) $\exists \ x \in N$, such that $x < 3$.

    (d) $\lim\limits_{n \to \infty} \dfrac{1}{n} = 0$.

    (e) He is a good player.

    (f) This sentence is true.

2. Find the solution set of the following sentences:

    (a) $x - 2 < 5$ and $x \in \{0, 1, 2, 3\}$,

    (b) $|x| + 1 < 3$ in $\{0, 1, 2, 3, 5\}$,

    (c) $(x + 2)(x - 1) = 0$ in $\{5, 6, 8\}$,

    (d) $2x^2 + 3x + 1 = 0$ in $\mathbb{Q}$,

    (e) $x^2 + 1 = 0$ in $\mathbb{R}$.

3. Determine the truth value of each of the following propositions:

    (a) For all real numbers x, $x^2 = 0$.

    (b) If $f(x) = x^3$, then $f'(x) = 3x^2$.

    (c) For any natural number n, $n^2 = n$.

    (d) 2 is a real number $\Leftrightarrow$ Damascus is in Syria.

    (e) There exists a natural number n such that $n^2 = n$.

    (f) There exists a rational number q such that $q < 2$.

    (g) e is a rational number and $\lim\limits_{n \to \infty} \dfrac{1}{n} = 1$.

    (h) For all $x \in \mathbb{R}$, $\sqrt{x^2} = |x|$ and $9 \neq 5$.

    (i) π is a rational number or π is a real number.

    (j) Paris is in France or $\sqrt{25} = 4$.

    (k) $\sim (\pi$ is not a rational number$)$.

    (l) $2 < 1 \Rightarrow 2 < 3$.

    (m) $3 > 7 \Rightarrow 10 < 8$.

    (n) 5 is real number $\Leftrightarrow$ 2/3 is natural number.

    (o) $2 = \sqrt{4} \Rightarrow \int x^2 dx = x^4$.

4. Express the following propositions with out using the symbol $\sim$:

    (a) $\sim (x < y)$,

    (b) $\sim (x > y)$,

    (c) $\sim (4 \leq x)$,

    (d) $\sim (y^3 \geq 2 + x)$.

5. Let P be "It is cold" and let Q be "It is raining". Give a simple verbal sentence which describes the following sentences:

    (a) $\sim P$,

    (b) $P \wedge Q$,

    (c) $P \vee Q$,

    (d) $Q \Leftrightarrow P$,

    (e) $P \Rightarrow \sim Q$,

    (f) $Q \vee \sim Q$,

    (g) $\sim Q \wedge \sim Q$,

    (h) $P \Leftrightarrow \sim Q$,

    (i) $\sim\sim Q$,

    (j) $(P \wedge \sim Q) \Rightarrow P$.

6. Determine the truth value of each of the following composite propositions:

    (a) If $3 + 2 = 7$, then $5 + 9 = 14$.

    (b) It is not true that $2 + 2 = 5$ if and only if $4 + 4 = 13$.

    (c) Paris is in England or London is in France.

    (d) It is not true that $1 + 1 = 3$ or $2 + 1 = 3$.

    (e) It is false that if Paris is in England then London is in France.

7. Let P, Q, R, and S be propositions. Find the truth tables for each of the following composite propositions:

    (a) $\sim P \wedge Q$,

    (b) $\sim (P \Rightarrow \sim Q)$,

    (c) $\sim (P \wedge Q)$,

    (d) $\sim P \vee \sim Q$,

    (e) $P \wedge P$,

    (f) $\sim (P \wedge Q)$,

    (g) $(P \vee \sim Q) \wedge R$,

    (h) $(P \wedge Q) \vee (R \wedge \sim S)$,

    (i) $(P \wedge Q) \Rightarrow (P \vee Q)$,

    (j) $\sim (P \wedge Q) \vee \sim (Q \Leftrightarrow P)$,

    (k) $(P \Rightarrow Q) \vee \sim (P \Leftrightarrow \sim Q)$,

    (l) $[P \Rightarrow (\sim Q \vee R)] \wedge \sim [Q \vee (P \Leftrightarrow \sim R)]$.

1.3 TAUTOLOGY AND CONTRADICTION

Definition 1.3.1 A compound proposition S is a tautology if S is true for any original propositions. In other words, a tautology will contain only T in the last column of its truth table.

Example 1.3.2 The proposition "P or not P", i.e. $P \vee \sim P$, is a tautology. This fact is verified by constructing a truth table.

P	$\sim P$	$P \vee \sim Q$
T	F	T
F	T	T

Definition 1.3.3 A compound proposition S is a contradiction if S is false for any original propositions. In other words, a contradiction will contain only F in the last column of its truth table.

Example 1.3.4 The proposition "P and not P", i.e. $P \wedge \sim P$, is a contradiction. The fact is verified by the following table:

P	$\sim P$	$P \wedge \sim Q$
T	F	F
F	T	F

Remark 1.3.5 Let P and Q be propositions. Then

(a) $(P \equiv Q)$ if and only if $(P \Leftrightarrow Q)$ is a tautology,

(b) P is a contradiction if and only if $\sim P$ is a tautology.

Definition 1.3.6 The proposition $Q \Rightarrow P$ is called the converse of $P \Rightarrow Q$.

Definition 1.3.7 The proposition $\sim P \Rightarrow \sim Q$ is called the inverse of $P \Rightarrow Q$.

Definition 1.3.8 The proposition $\sim Q \Rightarrow \sim P$ is called the contrapositive of $P \Rightarrow Q$.

Exercises

1. Show that $(P \Rightarrow Q) \not\equiv (Q \Rightarrow P)$.

2. Show that $(P \Rightarrow Q) \equiv (\sim Q \Rightarrow \sim P)$.

3. The inverse of $P \Rightarrow Q$ is $\sim P \Rightarrow \sim Q$. Show $(P \Rightarrow Q) \not\equiv (\sim P \Rightarrow \sim Q)$.

4. Show that $P \Rightarrow Q$ iff $P \wedge \sim Q$ is a contradiction.

5. Put the proposition

   "Every differentiable function at a point is continuous at that point" in form $P \Rightarrow Q$, and write its inverse and its contrapositive.

6. Prove the (Algebra of propositions):

1. Idempotent laws. Let P be a proposition. Then

   (a) $P \vee P \equiv P$,
   (b) $P \wedge P \equiv P$.

2. Associative laws. Let P, Q, and R be propositions. Then

   (a) $(P \vee Q) \vee R \equiv P \vee (Q \vee R)$,
   (b) $(P \wedge Q) \wedge R \equiv P \wedge (Q \wedge R)$.

3. Commutative laws. Let P and Q be propositions. Then

   (a) $P \vee Q \equiv Q \vee P$,
   (b) $P \wedge Q \equiv Q \wedge P$.

4. Distribution laws. Let P, Q, and R be propositions. Then

   (a) $P \vee (Q \wedge R) \equiv (P \vee Q) \wedge (P \vee R)$,
   (b) $P \wedge (Q \wedge R) \equiv (P \wedge Q) \vee (P \wedge R)$.

5. Identity laws. Let P be a proposition. Then

   (a) $P \vee 0 \equiv P$,
   (b) $P \wedge I \equiv P$,
   (c) $P \vee I \equiv I$,
   (d) $P \wedge 0 \equiv 0$.

where I represents tautology and 0 represents contradiction.

6. Complement laws. Let P be a proposition. Then

   (a) $P \vee \sim P \equiv I$,
   (b) $P \wedge \sim P \equiv 0$,
   (c) $\sim (\sim P) \equiv P$,
   (d) $\sim I \equiv 0$ and $\sim 0 \equiv I$.

7. De Morgan's laws. Let P and Q be propositions. Then

   (a) $\sim (P \wedge Q) \equiv \sim P \vee \sim Q$,
   (b) $\sim (P \wedge Q) \equiv \sim P \wedge \sim Q$.

We prove, for example, that the distribution law 4(b) holds. For this purpose, we compose the following table:

P	Q	R	$Q \vee R$	$P \wedge (Q \vee R)$	$P \wedge Q$	$P \wedge R$	$(P \wedge Q) \vee (P \wedge R)$
T	T	T	T	T	T	T	T
T	T	F	T	T	T	F	T
T	F	T	T	T	F	T	T
T	F	F	F	F	F	F	F
F	T	T	T	F	F	F	F
F	T	F	T	F	F	F	F
F	F	T	T	F	F	F	F
F	F	F	F	F	F	F	F

From this table we see that 4.(b) holds.

7. Prove that $P \Rightarrow (Q \wedge R) \equiv (P \Rightarrow Q) \wedge (P \Rightarrow R)$.

8. Prove that $P \vee Q \equiv \sim (\sim P \wedge \sim Q)$.

9. Use Algebra of propositions (or law 6) to show that:

(a) $(P \vee Q) \wedge \sim P \equiv \sim P \wedge Q$,
(b) $\sim (P \vee Q) \vee (\sim P \wedge Q) \equiv \sim P$,
(c) $\sim (P \vee \sim Q) \equiv \sim P \wedge Q$.

10. Determine which of the following is a tautology:

(a) $P \Rightarrow P \wedge Q$,
(b) $P \Rightarrow P \vee Q$,
(c) $(P \wedge Q) \Rightarrow (P \Leftrightarrow Q)$.

11. Find the truth table of each of the following propositions:

(a) $\sim P \wedge \sim Q$,
(b) $\sim (\sim P \Leftrightarrow Q)$,
(c) $P \Rightarrow (\sim P \vee Q)$,
(d) $(P \wedge \sim Q) \Rightarrow (\sim P \vee Q)$.

12. Prove (a) $(P \Rightarrow \sim Q) \equiv (Q \Rightarrow \sim P)$,

(b) $[(P \wedge Q) \Rightarrow R] \equiv (P \Rightarrow R) \vee (Q \Rightarrow R)$,
(c) $[(P \Rightarrow Q) \Rightarrow R] \equiv [(P \wedge \sim R) \Rightarrow \sim Q]$.

1.4   QUANTIFIERS

Let A be a set. An open sentence on A is an expression, denoted by $P(x)$, which has the property that $P(a)$ is true or false for each $a \in A$.

Example 1.4.1 Let $P(x)$ be "$x + 3 > 8$". Then $P(x)$ is an open sentence on N.

Example 1.4.2 Let $P(x)$ be "$x + 3 > 8$". Then $P(x)$ is not an open sentence on C since inequalities are not defined for all complex numbers.

If $P(x)$ is an open sentence on a set A, then the set of elements $a \in A$ with the property that $P(a)$ is true, is called truth set T_p of $P(x)$. In other words,

$$T_p = \{x : x \in A, P(x) \text{ is true } \},$$

or, simply,

$$T_p = \{x : P(x)\}.$$

Example 1.4.3 Let $P(x)$ be "$x + 3 > 8$" defined on $\mathbb{N}$. Then the truth set of $P(x)$ on N is

$$T_p = \{x : x \in \mathbb{N}, x + 3 > 8\} = \{6, 7, 8, 9, \ldots\}.$$

Example 1.4.4 Let $P(x)$ be "$x + 2 < 1$", defined on $\mathbb{N}$. Then

$$T_p = \{x; x \in \mathbb{N}, x + 2 < 1\} = \phi.$$

Example 1.4.5 Let $P(x)$ be "$x + 3 \geq 4$", defined on $\mathbb{N}$. Then

$$T_p = \{x : x \in \mathbb{N}, x + 3 \geq 4\} = \{1, 2, 3, \ldots\} = N.$$

Definition 1.4.6 Let $P(x)$ be an open sentence on a set A. Then the proposition $(\forall\, x \in A)$ or $\forall\, x\; P(x)$ can be read as "For every element x in A, $P(x)$ holds" or simply, "For all x, $P(x)$".

The symbol $\forall$, which reads "for all" or "for every", is called the universal quantifier.

Example 1.4.7 The proposition for all natural number n, $n+3 \geq 4$ is true since

$$\{n : n+3 \geq 4\} = \{1,2,3,\ldots\} = N.$$

Example 1.4.8 The proposition for all natural number n, $n+2 > 5$ is false since

$$\{n : n+2 > 5\} = \{4,5,6,\ldots\} \neq \mathbb{N}.$$

Definition 1.4.9 Let $P(x)$ be an open sentence on a set A. Then the proposition "$\exists\, x \in A)\; P(x)$" or "$\exists\, x,\; P(x)$" reads "There exists $x \in A$ such that $P(x)$ holds" or simply, "For some x, $P(x)$".

The symbol $\exists$, which reads "there exists" or "for some" or "for at least", is called the existential quantifier.

Example 1.4.10 The proposition exists $n \in \mathbb{N}$, $n+4 < 7$ is true since

$$\{n \in N : n+4 < 7\} = \{1,2\} \neq \phi.$$

Example 1.4.11 The proposition exists $n \in \mathbb{N}$, $n+6 < 5$ is false since

$$\{n \in N : n+6 < 5\} = \phi.$$

Definition 1.4.12 For an open sentence $P(x)$, the proposition "$(\exists\,!x),\; P(x)$" is read "There exists a unique x, such that $P(x)$". The sentence "$(\exists\,!x),\; P(x)$" is true, when the truth set for $P(x)$ contains exactly one element from the universe; hence $\exists\,!$ is called the unique existence quantifier.

Definition 1.4.13 Now we define the negation of propositions, which contain quantifiers. The negation of the proposition "All the integers are positive" reads "It is not true that all integers are positive"; in other words, there exists at least one integer, which is not positive.

Symbolically, if $\mathbb{Z}$ denotes the set of integer, then the above can be written as

$$\sim (\forall\, x \in \mathbb{Z})(x \text{ is positive}) \equiv (\exists\, x \in \mathbb{Z})(x \text{ is not positive}).$$

Furthermore, if $P(x)$ denotes "x is positive", then the above can be written as

$$\sim (\forall\, x \in \mathbb{Z})P(x) \equiv (\exists\, x \in Z) \sim P(x).$$

Theorem 1.4.14 (De Morgan) If $P(x)$ is an open sentence, then

$$(a) \sim (\forall\, x)P(x) \equiv (\exists\, x) \sim P(x),$$
$$(b) \sim (\exists\, x)P(x) \equiv (\forall\, x) \sim P(x).$$

Proof (a) The sentence $\sim (\forall x)P(x)$ is true

$$\text{iff } (\forall x)P(x) \text{ is false,}$$
$$\text{iff the truth set of } P(x) \text{ is not the universe,}$$
$$\text{iff the truth set of } \sim P(x) \text{ is not empty,}$$
$$\text{iff } (\exists x) \sim P(x) \text{ is true.}$$

Thus, $\sim (\forall x)P(x)$ is true if and only if $(\exists x) \sim P(x)$ is true. So the proposition is true.

(b) See Exercise 1.

Example 1.4.15 The negation of the proposition

"For all natural numbers $n, n+3 > 7$"

is equivalent to the proposition

"There exists an n such that $n+3 \leq 7$".

In other words,

$$\sim (\forall n \in \mathbb{N})(n+3 > 7) \equiv (\exists n \in \mathbb{N})(n+3 \leq 7).$$

Example 1.4.16 The negation of the proposition "All prime numbers are odd" is equivalent to the proposition "there exists a number that is prime and not odd".

In other words, if A is the set of prime numbers, then

$$\sim (\forall x \in A)(x \text{ is odd}) \equiv (\exists x \in A)(x \text{ is not odd}).$$

Exercises

1. Complete the proof of Theorem 1.4.14.

2. Translate the following English sentences into symbolic sentences with quantifiers:

   (a) There exists a rational number between arbitrary two different real numbers.

   (b) The set of real numbers contains a smallest positive integer.

   (c) For every positive real number x, there is a unique real number y, such that $2y = x$.

   (d) For every nonzero complex number, there is a unique complex number, such that their product is π.

   (e) For every complex number, there is at least one complex number, such that the product of the two complex numbers is a real number.

3. Which of the following propositions are true? (Here the universal set is R.)

(a) $(\forall x),\ |x| = x$.

(b) $(\forall x),\ x^2 = x$.

(c) $(\forall x),\ x+1 > x$.

(d) $(\exists x),\ x+2 = x$.

(e) $(\exists x),\ |x| = 0$.

4. Negate each of the propositions in Exercise 3.

5. Let $A = \{1,2,3,4,5\}$. Determine the truth value of each of the propositions.

(a) $(\exists x \in A)\ x+3 = 10$.

(b) $(\forall x \in A)\ x+3 < 10$.

(c) $(\exists x \in A)\ x+3 < 5$.

(d) $(\forall x \in A)\ x+3 \le 7$.

6. Negate each of the propositions in Exercise 5.

7. Let $\{1,2,3\}$ be the universal set. Determine the truth value of each of the following propositions:

(a) $(\exists x)\ (\forall y),\ x^2 < y+1$.

(b) $(\forall x)\ (\forall y),\ x^2 + y^2 < 12$.

(c) $(\forall x)\ (\forall y),\ x^2 + y^2 < 12$.

(d) $(\exists x)\ (\forall y)\ (\exists z),\ x^2 + y^2 < 2z^2$.

(e) $(\exists x)\ (\exists y)\ (\forall z),\ x^2 + y^2 < 2z^2$.

8. Negate each of the following propositions:

(a) $(\forall x)\ (\exists y)\ P(x) \vee Q(x)$.

(b) $(\exists x)\ (\forall y)\ P(x) \Rightarrow Q(x)$.

(c) $(\exists x)\ (\exists y)\ P(x) \wedge Q(x)$.

9. Which of the following are true?

(a) $(\forall x \in N)\ x+x \ge x$.

(b) $(\forall x \in R)\ x+x \ge x$.

(c) $(\exists x \in N)\ 2x+3 = 6x+7$.

(d) $(\exists x \in R)\ 3^x = x^2$.

(e) $(\exists x \in R)\ 3^x = x$.

(f) $(\forall x \in R)\ x^2 + 6x + 5 \ge 0$.

(g) $(\forall x \in R)\ x^2 + 4x + 5 \ge 0$.

(h) $(\exists\, x \in \mathbb{N})\, x^2 + x + 41$ is prime.

(i) $(\forall\, x \in \mathbb{N})\, x^2 + x + 41$ is prime.

10. Which of the following are true? (Here the universal set is the set of real numbers).

(a) $(\forall\, x)\, (\exists\, y)\, x + y = 0.$

(b) $(\exists\, x)\, (\forall\, y)\, x + y = 0.$

(c) $(\exists\, x)\, (\forall\, y)\, x^2 + y^2 = -1.$

(d) $(\forall\, x)\, x > 0 \Rightarrow (\exists\, y)\, y < 0 \wedge xy > 0.$

(e) $(\forall\, y)\, (\exists\, x)\, (\forall\, z)\, xy = xz.$

(f) $(\exists\, !y)\, y < 0 \wedge y + 3 = 0.$

(g) $(\forall\, y)\, (\exists\, !x)\, x = y^2.$

1.5   LOGICAL REASONING

Definition 1.5.1 An argument is an assertion that a given set of propositions $S_1, S_2, \ldots, S_n$, called premises, yields another proposition S, called the conclusion. Such an argument will be denoted by $S_1, S_2, \ldots, S_n \vdash S$.

Note that an argument is a proposition and therefore has a truth value. If an argument is true, it is called a valid argument; if not, it is called a fallacy.

Example 1.5.2

1. Consider the following propositions:

   S_1 : Some mathematicians are philosophers,
   S_2 : Ahmad is a mathematician,
   S : Then Ahmad is a philosopher.

   The argument S_1, $S_2 \vdash S$ does not valid, because not all mathematicians are philosophers.

2. Consider the following propositions:

   S_1 : All poets are interesting people,
   S_2 : Ahmad is an interesting person,
   S : Ahmad is a poet.

   The argument S_1, $S_2 \vdash S$ is fallacy.

3. Consider the following propositions:

   S_1 : No college professor is wealthy,
   S_2 : Some poets are wealthy,
   S : Some poets are not college professors.

   The argument S_1, $S_2 \vdash S$ is valid.

4. Consider the following propositions:

   S_1 : If it rains, Nazar will be sick,
   S_2 : Nazar was not sick,
   S : It did not rain.
   The argument S_1, $S_2 \vdash S$ is valid.

Remark 1.5.3

1. Note that the truth value of an argument

$$S_1, S_2, \ldots, S_n \vdash S$$

does not depend upon the particular truth value of each of the propositions in the argument.

2. Note that the argument

$$S_1, S_2, \ldots, S_n \vdash S$$

is valid if and only if the proposition

$$(S_1 \wedge S_2 \wedge \cdots \wedge S_n) \Rightarrow S \text{ is a tautology.}$$

3. Note that Venn diagrams are very often used to determine the validity of an argument.

Example 1.5.4 Let the propositions be:

S_1 : Some students are lazy,
S_2 : All males are lazy,

S : Some students are males.

Show that the argument $S_1, S_2 \vdash S$ does not valid by constructing a Venn diagram, in which the premises S_1 and S_2 hold, but the conclusion S does not hold.

Solution: Consider the following diagram:

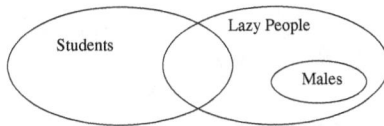

Notice that both premises hold, but the conclusion does not hold.

Example 1.5.5 Let the propositions be:

S_1 : All students are lazy.
S_2 : Nobody who is wealthy is a student.

S : Lazy people are not wealthy.

Show that the argument $S_1, S_2 \vdash S$ does not valid by constructing a Venn diagram, in which the premises S_1 and S_2 hold, but the conclusion S does not hold.

Solution: Consider the following Venn diagram:

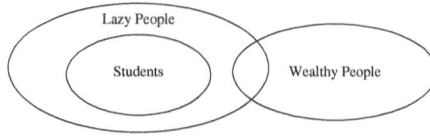

Notice that the premises hold, but the conclusion does not hold. Hence the argument is not valid.

Example 1.5.6 Determine the validity of the following argument:

$$P \Rightarrow Q, \sim P \vdash \sim Q.$$

Solution: Construct the necessary truth tables.

P	Q	$P \Rightarrow Q$	$(P \Rightarrow Q) \wedge \sim P$	$[(P \Rightarrow Q) \wedge \sim P] \Rightarrow \sim Q$
T	T	T	F	T
T	F	F	F	T
F	T	T	T	F
F	F	T	T	T

Since $[(P \Rightarrow Q) \wedge \sim P] \Rightarrow \sim Q$ is not a tautology, then the argument is fallacy.

Example 1.5.7 Prove that the argument $P \Rightarrow \sim Q, R \Rightarrow Q, R \vdash \sim P$ is valid.

Solution: Construct the following truth tables:

P	Q	R	$P \Rightarrow \sim Q$	$R \Rightarrow Q$	$\sim P$	$[(P \Rightarrow \sim Q) \wedge (R \Rightarrow Q) \wedge R] \Rightarrow \sim P$
T	T	T	F	T	F	T
T	T	F	F	T	F	T
T	F	T	T	F	F	T
T	F	F	T	T	F	T
F	T	T	T	T	T	T
F	T	F	T	T	T	T
F	F	T	T	F	T	T
F	F	F	T	T	T	T

Since the proposition $[(P \Rightarrow \sim Q) \wedge (R \Rightarrow Q) \wedge R] \Rightarrow \sim P$ is a tautology, then the argument $P \Rightarrow \sim Q, R \Rightarrow Q, R \vdash \sim P$ is valid.

1.6   MATHEMATICAL PROOF

1.6.1   DIRECT AND CONTRAPOSITIVE PROOF

Definition 1.6.1 Let $S_1, S_2, \ldots, S_n$ be a set of propositions yielding another proposition S. If the argument

$$S_1, S_2, \ldots, S_n \vdash S$$

is valid, then it is called a proof.

The first method of proof, which we will examine, is the direct proof of a conditional sentence. How do we prove a statement in the form $P \Rightarrow Q$?

This implication is false only when P is true and Q is false. A direct proof of $P \Rightarrow Q$ will have the following form:

Proof.

$$\text{Assume } P$$

$$\vdots$$

$$\vdots$$

$$\text{Therefore } Q$$

$$\text{Thus } P \Rightarrow Q.$$

Example 1.6.2 Prove that if x is even integer, then x^2 is even.

Proof Suppose x is an even integer
  Then $x = 2k$ for some integer k.
  Thus $x^2 = 4k^2$
  $\qquad\quad = 2(2k^2)$
  $\qquad\quad = 2t$ for some integer t.
Since x^2 is twice the integer t, x^2 is even.

Remark 1.6.3 In the above proof, we used the tautology

$$[(P \Rightarrow S_1) \wedge (S_1 \Rightarrow S_2) \wedge \ldots (S_n \Rightarrow R)] \Rightarrow (P \Rightarrow R),$$

where

$$P : x \text{ is even,}$$
$$S_1 : x = 2k,$$
$$S_2 : x^2 = 4k^2,$$
$$R : x^2 \text{ is even.}$$

A second form of proof for a conditional sentence is a proof by contrapositive.
    The idea here is that since $P \Rightarrow Q \equiv\, \sim Q \Rightarrow\, \sim P$, we first give a direct proof of $\sim Q \Rightarrow\, \sim P$ and then conclude $P \Rightarrow Q$. That is, instead of proving $P \Rightarrow Q$, we prove its equivalent statement $\sim Q \Rightarrow\, \sim P$. Such a form of proof by contrapositive is as follows:

Proof Suppose $\sim Q$.
  $\vdots$

  $\vdots$
  Therefore $\sim P$ (Via a direct proof).
  Thus $\sim Q \Rightarrow\, \sim P$.
  Therefore $P \Rightarrow Q$.

Example 1.6.4 Prove that a^2 is even $\Rightarrow a$ is even.

Proof Consider the contrapositive

$$\sim (a \text{ is even}) \Rightarrow\, \sim (a^2 \text{ is even}).$$

That is,

$$a \text{ is odd} \Rightarrow a^2 \text{ is odd.}$$

Since a is odd, then

$$a = 2k + 1, \text{ for some integer } k.$$

Then

$$a^2 = (2k+1)^2$$
$$= 4k^2 + 4k + 1$$
$$= 2(2k^2 + 2k) + 1$$
$$= 2r + 1, \text{ where } r \text{ is an integer.}$$

Thus a^2 is odd.
So, by contrapositive, if a^2 is even, then a is even.

1.6.2 CONTRADICTION PROOF

A third form of proof is by contradiction, making use of the tautology

$$P \Leftrightarrow [\sim P \Rightarrow (Q \wedge \sim Q)].$$

To prove a proposition P, it is sufficient to prove that

$$\sim P \Rightarrow (Q \wedge \sim Q).$$

A proof by contradiction has the following form:
Proof Suppose $\sim P$.

$$\vdots$$

Therefore Q.

$$\vdots$$

Therefore $\sim Q$.
Hence $Q \wedge \sim Q$, a contradiction.
Thus P.

Example 1.6.5 Prove that $x \neq 0 \Rightarrow x^{-1} \neq 0$ as rational numbers.

Proof Consider $P : x \neq 0$,
$\qquad\qquad Q : x^{-1} \neq 0$.
We need to show that $P \Rightarrow Q$.
Suppose $\sim (P \Rightarrow Q)$ is true.
Since $\sim (P \Rightarrow Q) \equiv P \wedge \sim Q$,
then $P \wedge \sim Q$ is true, that is $x \neq 0 \wedge x^{-1} = 0$ is true.
Since $x \cdot x^{-1} = 1$ and $x^{-1} = 0 \Rightarrow x \cdot x^{-1} = x \cdot 0 = 0$,
then $1 = 0$.
So, one has $(1 = 0)$ and $(1 \neq 0)$.
It contradicts the number equality. Consequently $\sim (P \Rightarrow Q)$ is false.
Thus $P \Rightarrow Q$ is true.
Hence $x \neq 0 \Rightarrow x^{-1} \neq 0$.

Example 1.6.6 Prove that $\sqrt{2}$ is an irrational number.

Proof Suppose that $\sqrt{2}$ is a rational number.
   Then $\sqrt{2} = \frac{a}{b}$, where a, b are positive integers with $gcd(a, b) = 1$.
Hence we obtain

$$\sqrt{2} = \frac{a}{b} \Rightarrow a^2 = 2b^2$$
$$\Rightarrow a^2 \text{ is even}$$
$$\Rightarrow a \text{ is even}$$
$$\Rightarrow a = 2k, \text{ for some integer } k.$$

Then

$$a^2 = 2b^2 \Rightarrow (2k)^2 = 2b^2$$
$$\Rightarrow 4k^2 = 2b^2$$
$$\Rightarrow 2k^2 = b^2$$
$$\Rightarrow b^2 \text{ is even}$$
$$\Rightarrow b \text{ is even}$$
$$\Rightarrow b = 2t, \text{ for some integer } t.$$

Thus 2 divides $gcd(a, b)$ and hence $gcd(a, b) \geq 2$.
The contradiction is $gcd(a, b) = 1$ and $gcd(a, b) = 2$.
We conclude that $\sqrt{2}$ is an irrational number.

Exercises

1. For each set of premises, find a conclusion, such that the argument is valid, and such that the premise is necessary for the conclusion.

   (a) $S_1, S_2, S_3 \vdash S$,

   S_1 : No student is lazy,
   S_2 : Jallal is an artist,
   S_3 : All artists are lazy,
   S : _____.

   (b) $S_1, S_2, S_3, S_4 \vdash S$,

   S_1 : All lawyers are wealthy,
   S_2 : Poets are temperamental,
   S_3 : Ahmad is a lawyer,
   S_4 : No temperamental person is wealthy,
   S : _____.

2. Determine the validity of the following argument:

$$P \Rightarrow Q, Q \vdash P$$

3. Determine the validity of the following argument:

S_1 : If it rains, Anwar will be sick,
S_2 : It did not rain,

___

S : Anwar was not sick.

4. Determine the validity of the following argument for the proposed conclusion:

S_1 : All poets are poor,
S_2 : In order to be a teacher, one must graduate from college,
S_3 : Some mathematicians,
S_4 : No college graduate is poor.

___

S : Teachers are not poor.

5. Determine the validity of the following argument for the proposed conclusion.:

S_1 : All mathematicians are interesting people,
S_2 : Some teachers sell insurance,
S_3 : Some philosophers are mathematicians,
S_4 : Only uninteresting people become insurance salesman.

___

S : Some teachers are not philosophers.

6. Determine the validity of the following arguments:

(a) $P \Rightarrow Q, R \Rightarrow \sim Q \vdash \sim P$.

(b) $P \Rightarrow Q, \sim R \Rightarrow \sim Q \vdash P \Rightarrow \sim R$.

7. For the given premises, determine a suitable conclusion so that the argument is valid.

(a) $P \Rightarrow \sim Q, Q$.
(b) $P \Rightarrow \sim Q, R \Rightarrow Q$.
(c) $P \Rightarrow \sim Q, \sim P \Rightarrow R$.
(d) $P \Rightarrow \sim Q, R \Rightarrow P, Q$.

8. Prove that if a, b are even integers, then $a + b$ is even.

9. Prove that if a is an even integer and b is an odd integer, then $a + b$ is odd.

10. Prove that x is odd integer iff $x + 1$ is even integer.

11. Prove that if x is a real number, then $|x| \geq 0$.

12. Prove that if x, y are real numbers, then $|xy| = |x||y|$.

13. Prove that $\forall x \in R, x \leq |x|$.

14. Prove that $\exists\, x \in R, x^2 = x$.

15. Prove that $\exists\, y \in R, \forall\, x \in R, x + y = x$.

16. Prove that $\exists\, y \in R, \forall\, x \in R, xy = x$.

17. Prove that
$$\forall\, x[P(x) \wedge Q(x)] \Leftrightarrow [\forall\, xP(x) \wedge \forall\, xQ(x)].$$

18. Let $A = \{2,3\}$. Prove that $\forall\, x \in A\ (1 < x \Rightarrow 1 < x^2)$.

19. Prove that $\forall\, x \in N\ (1 < x \Rightarrow 1 < x^2)$.

20. Prove that the polynomial $f(x) = x - 3$ has a unique zero.

21. Prove that every nonzero real number has a unique multiplicative inverse.

22. Find the contrapositive of each of the following propositions:

    (a) If he has courage, he will win.
    (b) It is necessary to be strong, in order to be sailor.
    (c) Only if he does not tire, he will win.
    (d) It is sufficient for it to be a square, in order to be a rectangle.

23. Determine the following contrapositive of the

    (a) Contrapositive of $P \Rightarrow Q$,
    (b) Converse of $P \Rightarrow Q$,
    (c) Inverse of $P \Rightarrow Q$.

24. Prove that if x is a rational number and y is an irrational number, then $x + y$ is an irrational number.

25. Prove that if x is an integer, then $(x^2 - x)$ is an even integer.

26. Prove that if $f(x) = f(x + p)$ for all $p > 0$, then f is constant function.

1.7 INDUCTION

1.7.1 INTRODUCTION

Proof by induction involves propositions, which depend on the natural numbers, $n = 1, 2, 3, \ldots$. It often uses summation notation, which we now briefly review before discussing induction itself.

We use the symbol $\sum$ to denote a sum over its argument for each natural number i from the lowest value for i (appears below) to the maximum value for i (appears above).

For example, we write the sum of natural numbers up to a value n as:

$$1 + 2 + 3 + \cdots + (n - 1) + n = \sum_{i=1}^{i}.$$

Example 1.7.1 Write the following sums without the summation notation:

(a) $\displaystyle\sum_{i=1}^{4} \frac{1}{3^i}$,

(b) $\displaystyle\sum_{i=1}^{3} (2i+1)$,

(c) $\displaystyle\sum_{i=1}^{5} \frac{i-1}{i}$.

The above sums, written out without summation notation, are:

(a) $\displaystyle\sum_{i=1}^{4} \frac{1}{3^i} = \frac{1}{3} + \frac{1}{9} + \frac{1}{27} + \frac{1}{81}$,

(b) $\displaystyle\sum_{i=1}^{3} (2i+1) = (2 \times 1 + 1) + (2 \times 2 + 1) + (2 \times 3 + 1) = 3 + 5 + 7$,

(c) $\displaystyle\sum_{i=1}^{5} \frac{i-1}{i} = \frac{1-1}{1} + \frac{2-1}{2} + \frac{3-1}{3} + \frac{4-1}{4} + \frac{5-1}{5} = 0 + \frac{1}{2} + \frac{2}{3} + \frac{3}{4} + \frac{4}{5}$.

1.7.2   INDUCTION PRINCIPLE

In Section 1.6, we have studied some types of mathematical proofs. Here we discuss another proof technique, called mathematical induction. This technique will be used to prove a variety of mathematical propositions.

We illustrate the induction principle by the following:

Let $P(n)$ be a proposition, which involves a natural numbers n, i.e., $n = 1, 2, 3, \ldots$. Then $P(n)$ is true for all n if and only if

(a) $P(1)$ is true, and

(b) $P(k) \Rightarrow P(k+1)$ for all natural number k.

Example 1.7.2 Prove that for all positive integers n,

$$1 + 2 + 3 + \cdots + n = \frac{n(n+1)}{2},$$

i.e., $P(n) = \displaystyle\sum_{i=1}^{n} i = \frac{n(n+1)}{2}$.

Solution:

Step (a) Let us check whether the formula is true for $n = 1$. Indeed

- $\sum_{i=1}^{1} i = 1$,
- $\frac{1(1+1)}{2} = \frac{1(2)}{2} = \frac{2}{2} = 1$,

i.e., the formula is true for $n = 1$.

Step (b) Now suppose that the formula is true for $n = k$ and prove that it remains true for $n = k + 1$. In other words, the question is:

$$\sum_{i=1}^{k} i = \frac{k(k+1)}{2} \xrightarrow{?} \sum_{i=1}^{k+1} i = \frac{(k+1)(k+2)}{2}.$$

The sum for $n = k + 1$ may be written as

$$\sum_{i=1}^{k+1} i = \sum_{i=1}^{k} i + (k+1)$$
$$= \frac{k(k+1)}{2} + (k+1)$$
$$= \frac{k(k+1) + 2(k+1)}{2}$$
$$= \frac{(k+1)(k+2)}{2}.$$

So, we have shown that if $p(k)$ is true, then $P(k+1)$ is true.

Example 1.7.3 Prove that for all positive integers n, the following results hold.

1. $\sum_{i=1}^{n} (2i - 1) = n^2$,

2. $\frac{1}{2} + \frac{1}{4} + \frac{1}{8} + \cdots + \frac{1}{2^n} = \frac{2^n - 1}{2^n}$.

Solution:

1. Step (a) For $n = 1$ the formula is true, since

   - $\sum_{i=1}^{1} (2i - 1) = 2 \cdot 1 - 1 = 2 - 1 = 1$,

   - $(1)^2 = 1$.

   Step (b) Assume that the result is true for $n = k$, i.e

   $$\sum_{i=1}^{k} (2i - 1) = k^2.$$

As for $n = k+1$ the sum may be written as

$$\sum_{i=1}^{k+1}(2i-1) = \sum_{i=1}^{k}(2i-1) + (2(k+1)-1)$$
$$= k^2 + 2k + 1$$
$$= (k+1)^2,$$

then we have shown that if $p(k)$ is true, then $P(k+1)$ is true.

2. Step (a) For $n = 1$ the formula is true, since

- $\frac{1}{2^1} = \frac{1}{2}$,

- $\frac{2^1 - 1}{2^1} = \frac{2-1}{2} = \frac{1}{2}$.

Step (b) Assume the result is true for $n = k$, i.e.

$$\frac{1}{2} + \frac{1}{4} + \frac{1}{8} + \cdots + \frac{1}{2^k} = \frac{2^k - 1}{2^k}$$

As for $n = k+1$ the sum may be written as

$$\frac{1}{2} + \frac{1}{4} + \frac{1}{8} + \cdots + \frac{1}{2^k} + \frac{1}{2^{k+1}} = \frac{2^k - 1}{2^k} + \frac{1}{2^{k+1}}$$
$$= \frac{2^k - 1}{2^k} \cdot \frac{2}{2} + \frac{1}{2^{k+1}}$$
$$= \frac{2^{k+1} - 2}{2^{k+1}} + \frac{1}{2^{k+1}}$$
$$= \frac{2^{k+1} - 1}{2^{k+1}}.$$

then we have shown that if $p(k)$ is true, then $P(k+1)$ is true.

Example 1.7.4 For a natural number n prove the following propositions:

(1) If $n \geq 2$, then $n^3 - n$ is divisible by 3.

(2) $n < 2^n$.

Solution:

1. Step (a) for $n = 2$,

$$2 \geq 2.$$
$$n^3 - n = (2)^3 - 2 = 6 = 3 \times 2, \text{ so divisible by 3.}$$

> An integer n is divisible by 3 if and only if there
> exists an integer r such that $n = 3r$.

Step (b) Assume the result is true for $n = k$, i.e. is

$$(k^3 - k = 3r) \xrightarrow{?} (k+1)^3 - (k+1) = 3r,$$

we have

$$(k+1)^3 - (k+1) = k^3 + 3k^2 + 3k + 1 - (k+1)$$
$$= (k^3 - k) + 3k^2 + 3k$$
$$= 3r + 3k^2 + 3k$$
$$= 3(r + k^2 + k)$$

The principle of induction thus implies that $n^3 - n$ is indeed divisible by 3 for all $n \geq 2$.

2. 1. Step (a) For $n = 1$,

$$n = 1.$$
$$2^n = 2^1. \ (1 < 2)$$

2. Step (b) Assume the result is true for $n = k$, i.e. is

$$k < 2^k \xrightarrow{?} k+1 < 2^{k+1}.$$

We have

$$k < 2^k \Rightarrow k+1 < 2^k + 1$$
$$\Rightarrow k+1 < 2^k + 2^k, \ (1 < 2^k)$$
$$\Rightarrow k+1 < 2 \times 2^k = 2^{k+1}$$

Hence $k+1 > 2^{k+1}$.

From the principle of induction thus implies that $n < 2^n$ natural number n.

Exercises

1. Write the first four terms for each given sequence.

   (a) $a_n = n - 1$,

   (b) $a_n = n + 2$,

   (c) $a_n = \frac{n-2}{n+1}$,

   (d) $a_n = (1 + \frac{2}{n})^n$,

   (e) $a_n = (-3)^{n-1}$,

   (f) $a_n = \frac{(-1)^{n+1}}{n2}$.

2. Find a general term a_n for the given sequence a_1, a_2, a_3, a_4

   (a) $-2, -1, 0, 1, \ldots$

   (b) $-3, 3, -3, 3, \ldots$

   (c) $2, \frac{3}{2}, \frac{4}{3}, \frac{5}{4}, \ldots$

   (d) $5, 25, 125, 625, \ldots$

   (e) $x, \frac{x^2}{2}, \frac{x^3}{3}, \frac{x^4}{4}, \ldots$

3. Expand the sums

   (a) $\displaystyle\sum_{i=1}^{3} (3i - 1),$

   (b) $\displaystyle\sum_{i=o}^{3} 3(2i + 1),$

   (c) $\displaystyle\sum_{j=1}^{3} \frac{1}{j^2}.$

4. Express the following in summation notation.

   (a) $1 + 20 + 400 + 8000.$

   (b) $-3 - 1 + 1 + 5 + 7.$

5. Use the principle of induction to prove the following results. Unless otherwise, assume n is a positive integer number.

   (a) $2 + 2^2 + 2^3 + \cdots + 2^n = 2^{n+1} - 2.$

   (b) $\displaystyle\sum_{i=1}^{n} i^2 = \frac{n(n+1)(2n+1)}{6}.$

   (c) $5^n - 1$ is divisible by 4.

   (d) $8^n + 3$ is divisible by 4.

   (e) $3^n > n^2.$

   (f) $\displaystyle\sum_{i=1}^{n} \frac{1}{i^2} \leq 2 - \frac{1}{n}.$

2 Set Theory

In this chapter, we shall discuss a basic language that is useful in describing all the branches of mathematics, namely the language of sets. We will use the word "set" to refer to any specified collection of objects. The objects in a given set are called elements (or members) of the set.

2.1 BASIC NOTIONS OF SET THEORY

In general, sets will be denoted by capital letters

$$A, B, C, D, E, \ldots.$$

The elements in our sets will be represented by lowercase letters

$$a, b, c, d, e \ldots.$$

If the object a is an element of set B (a an element belongs to a set B), we write $a \in B$, if not—that is, if

$$\sim (a \in B) - \text{ we write } a \notin B.$$

For example, if $\mathbb{Z}$ is the set of integers, we write $8 \in \mathbb{Z}$ and $5/6 \notin \mathbb{Z}$.

One way to describe a set is simply to list its elements between curly braces. For example, we can define the set A having exactly the four numbers 2, 4, 6, and 8 as elements by $A = \{2, 4, 6, 8\}$.

We call this the tabular form of a set. But if we define a particular set by stating properties that its elements must satisfy, for example let B be the set of all odd numbers, then we use a letter, usually x, to represent an arbitrary element and we write:

$$B = \{x : x \text{ is odd}\},$$

which reads "B is the set of numbers x such that x is odd". We call this the builder form of a set.

Definition 2.1.1
Let $\phi = \{x : x \neq x\}$. Then ϕ is a set with no elements and is called an empty set.

Example 2.1.2
$$\{x \in \mathbb{R} : x = x + 1\} = \phi = \{x \in \mathbb{N} : x < 0\}.$$

Definition 2.1.3
Let A and B be sets. We say that A is a subset of B iff every element of A is also an element of B. In symbols this is

$$A \subseteq B \Leftrightarrow (\forall x)(x \in A \Leftarrow x \in B).$$

DOI: 10.1201/9780429022838-2

Example 2.1.4

Let $A = \{1,3,4\}$, $B = \{-1,0,1,3,4\}$. Then $A \subset B$.

Example 2.1.5

$\mathbb{N} \subset \mathbb{R}, \mathbb{Z} \subset \mathbb{Q}$.

Theorem 2.1.6

(a) For any set A, ϕ is a subset of A.

(b) For any set A, $A \subseteq A$.

Proof

(a) Let A be any set. We need to show that

$$\forall x (x \in \phi \Rightarrow x \in A).$$

Which is equivalent to

$$x \notin A \Leftarrow x \notin \phi \text{ (contrapositive condition)}.$$

It is clear that $x \notin \phi$ (since ϕ is a set with no elements).
Then $x \notin A \Rightarrow x \notin \phi$.
That is, $x \in \phi \Rightarrow x \in A$.
Therefore $\phi \subseteq A$.

(b) The proof is left as Exercise 1.

Theorem 2.1.7

If A is a subset of B and B is a subset of C, then A is a subset of C. That is

$$(A \subseteq B \wedge B \subseteq C) \Rightarrow A \subseteq C.$$

Proof

To prove that $A \subseteq C$, we must show that

$$\forall x (x \in A \Rightarrow x \in C).$$

Since $A \subseteq B$, then

$$\forall x (x \in A \Rightarrow x \in B),$$

and since $B \subseteq C$, then

$$\forall x (x \in B \Rightarrow x \in C).$$

Hence $\forall x (x \in A \Rightarrow x \in C)$.
Therefore $A \subseteq C$.

Remark 2.1.8

For a given set A, the subsets ϕ and A are called improper subsets of A, while any subset of A other than ϕ or A is called a proper subset.

Definition 2.1.9

For an arbitrary set A, the set of all the subsets of A is called the power set of A. We denote the power set of A by $P(A)$. Thus

$$P(A) = \{B : B \subseteq A\}.$$

Example 2.1.10

Let $A = \{4,7\}$. Then

$$P(A) = \{\phi, \{4\}, \{7\}, A\}.$$

Example 2.1.11

Let $M = \{a,b,c\}$. Then

$$P(M) = \{\phi, \{a\}, \{b\}, \{c\}, \{a,b\}, \{a,c\}, \{b,c\}, M\}.$$

Theorem 2.1.12

If A is a set with n elements, the power set $P(A)$ has 2^n elements.

Proof

Suppose A has n elements. We may write A as $A = \{x_1, x_2, \ldots, x_n\}$. To describe a subset B of A, we need to know for each $x_i \in A$ whether the element is in B. For each x_i, there are two possibilities ($x_i \in B$ or $x_i \notin B$), so there are $2 \cdot 2 \cdot 2 \cdots \ldots \cdots 2$ (n factors) different ways of i making a subset of A. Therefore, $P(A)$ has 2^n elements.

Definition 2.1.13

Let A and B be sets. Then $A = B$ iff $A \subseteq B$ and $B \subseteq A$.

Example 2.1.14

Let $A = \{x : x$ is a solution to $x^2 - 4 = 0\}$ and $B = \{2, -2\}$. Prove that $A = B$.

Proof

We must show that $A \subseteq B$ and $B \subseteq A$.

Let

$$t \in A \Rightarrow t^2 - 4 = 0$$
$$\Rightarrow (t-2)(t+2) = 0$$
$$\Rightarrow t = 2 \text{ or } t = -2,$$

so $t \in B$. This proves the inclusion

$$A \subseteq B. \tag{2.1}$$

By substitution we see that both 2 and -2 are solutions to

$$x^2 - 4 = 0.$$

Thus

$$B \subseteq A. \tag{2.2}$$

It follows from Equations (2.1) and (2.2) that $A = B$.

Definition 2.1.15
All the sets under investigation will likely be subsets of a fixed set. We call this set the universal set or universe of discourse. We denote this set by U.

2.2   ELEMENTARY PROPERTIES OF SETS

2.2.1   UNION OF SETS

Definition 2.2.1
If A and B are sets. The union of A and B, denoted by $A \cup B$, is the set defined by

$$A \cup B = \{x : x \in A \text{ or } x \in B\}.$$

That is, $x \in A \cup B \Leftrightarrow x \in A \vee x \in B$.

Example 2.2.2
In the Venn diagram, we have shaded $A \cup B$, that is the area of A and the area of B

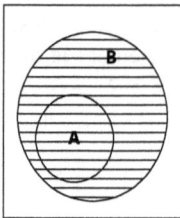

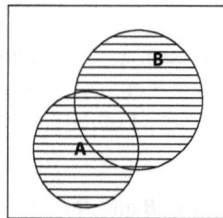

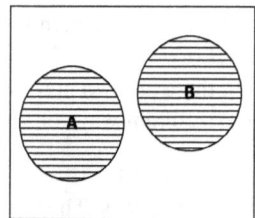

A ∪ B is shaded   A ∪ B is shaded   A ∪ B is shaded

Example 2.2.3
Let $X = \{a,b,c,d\}, Y = \{e,b,a,r\}$. Then

$$X \cup Y = \{a,b,c,d,e,r\}.$$

Example 2.2.4
Let $A = \{x \in \mathbb{N} : 3 < x < 10\}$ and $B = \{x \in \mathbb{N} : 1 \leq x \leq 3\}$.
Then $A \cup B = \{x \in \mathbb{N} : 1 \leq x < 10\}$.

Example 2.2.5
Let $\mathbb{N}_e = \{x \in \mathbb{N} : x \text{ is an even number}\}$
      $= \{2,4,6,8,\ldots\}$,
$\mathbb{N}_o = \{x \in \mathbb{N} : x \text{ is an odd number}\}$
      $= \{1,3,5,7,\ldots\}$.
Then

$$\mathbb{N}_e \cup \mathbb{N}_o = \{1,2,3,4,5,6,\ldots\} = \mathbb{N}.$$

Theorem 2.2.6
Let A and B be sets. Then

(a) $A \subseteq A \cup B$ and $B \subseteq A \cup B$,

(b) $A \subseteq B \Leftrightarrow A \cup B = B$.

Proof

(a) Let $x \in A$,
then

$$x \in A \Rightarrow x \in A \lor x \in B$$
$$\Rightarrow x \in A \cup B.$$

Thus $A \subseteq A \cup B$.
By the same manner, we prove that $B \subseteq A \cup B$.

(b) Suppose $A \subseteq B$ and $x \in A \cup B$.
Then

$$x \in A \cup B \Rightarrow x \in A \lor x \in B$$
$$\Rightarrow x \in B \lor x \in B \text{ (since } A \subseteq B)$$
$$\Rightarrow x \in B.$$

Thus $A \cup B \subseteq B$.
Since $B \subseteq A \cup B$ (by part (a)), then $A \cup B = B$.
Conversely, suppose that $A \cup B = B$.
Since $A \subseteq A \cup B$ (by part (a)), then $A \subseteq B$.

Theorem 2.2.7
Let A, B, and C be sets. Then

(a) $A \cup A = A$ (Idempotent law),

(b) $A \cup B = B \cup A$ (Commutative law),

(c) $A \cup (B \cup C) = (A \cup B) \cup C$ (Associative law).

Proof
We prove (b) and leave (a), (c) as Exercise 2. Suppose $x \in A \cup B$,
then

$$x \in A \cup B \Leftrightarrow x \in A \lor x \in B$$
$$\Leftrightarrow x \in B \lor x \in A.$$

Thus $A \cup B = B \cup A$.

Theorem 2.2.8
Let A be a set. Then

(a) $A \cup \phi = A$,

(b) $A \cup U = U$, where U is the universal set.

Proof

(a) Since $\phi \subseteq A$, then

$$A \cup \phi = A(\text{ by Theorem 2.2.6}).$$

(b) Since $A \subseteq U$, then

$$A \cup U = U(\text{ by Theorem 2.2.6}).$$

2.2.2 INTERSECTION OF SETS

Definition 2.2.9
If A and B are sets. The intersection of A and B, denoted by $A \cap B$, is defined by

$$A \cap B = \{x : x \in A \text{ and } x \in B\}.$$

That is,

$$x \in A \cap B \Leftrightarrow x \in A \wedge x \in B.$$

Example 2.2.10
In the Venn diagram, we have shaded $A \cap B$.

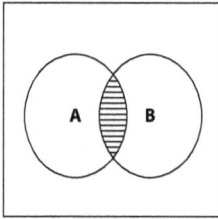

$A \cap B$ is shaded $A \cap B$ is shaded $A \cap B$ is shaded

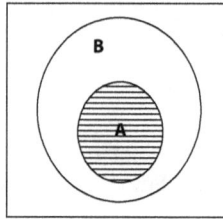

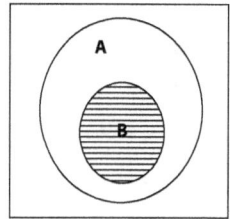

Example 2.2.11
Let $A = \{1, 2, x, y\}$, $B = \{2, 3, x, 4\}$. Then

$$A \cap B = \{2, x\}.$$

Example 2.2.12
Let $A = \{x \in \mathbb{N} : x \le 6\}$, $B = \{x : x \text{ is a prime number and } \le 6\}$.
Then

$$A \cap B = \{1, 2, 3, 4, 5, 6\} \cap \{2, 3, 5\}$$
$$= \{2, 3, 5\}.$$

Example 2.2.13
Let $A = \{x \in \mathbb{R} : 0 \le x \le 5\}$ and $B = \{x \in \mathbb{R} : 1/3 \le x \le 7\}$.
Then

$$A \cap B = \{x \in R : 1/3 \le x \le 5\}.$$

Definition 2.2.14
Let A, B be sets. We say that A and B are disjoint iff

$$A \cap B = \phi.$$

Example 2.2.15
Let $A = \{x \in \mathbb{N} : x \text{ is an odd number}\}$ and $B = \{x \in N : x \text{ is an even number}\}$.
   Since $A \cap B = \phi$, then A and B are disjoint. But $A \cap N \neq \phi$, thus A and N are not disjoint.

Theorem 2.2.16
Let A and B be sets. Then

(a) $A \cap B \subseteq A$ and $A \cap B \subseteq B$,

(b) $A \subseteq B \Leftrightarrow A \cap B = A$.

Proof

(a) To prove $A \cap B \subseteq A$, let $x \in A \cap B$. Then

$$x \in A \cap B \Rightarrow x \in A \wedge x \in B$$
$$\Rightarrow x \in A.$$

So, $A \cap B \subseteq A$
By the same manner, we prove that $A \cap B \subseteq B$.

(b) Assume that $A \subseteq B$ and let $x \in A$. Then

$$x \in A \Rightarrow x \in B.$$

Then $x \in A \wedge x \in B$.
That is

$$x \in A \Rightarrow x \in A \wedge x \in B$$
$$\Rightarrow x \in A \cap B.$$

Then $A \subseteq A \cap B$.
Since $A \cap B \subseteq A$ (by part (a)), then $A \cap B = A$.
Conversely, assume that $A \cap B = A$. As $A \cap B \subseteq B$ (by part (a)), then $A \subseteq B$.

Theorem 2.2.17
Let A, B, and C be sets. Then

(a) $A \cap A = A$ (Idempotent law),

(b) $A \cap B = B \cap A$ (Commutative law),

(c) $A \cap (B \cap C) = (A \cap B) \cap C$ (Associative law).

Proof

We prove (c) only, and leave (a), (c) as Exercise 3. Let $x \in A \cap (B \cap C)$. Then

$$x \in A \cap (B \cap C) \Leftrightarrow x \in A \wedge x \in (B \cap C)$$
$$\Leftrightarrow x \in A \wedge (x \in B \wedge x \in C)$$
$$\Leftrightarrow (x \in A \wedge x \in B) \wedge x \in C$$
$$\Leftrightarrow x \in (A \cap B) \wedge x \in C$$
$$\Leftrightarrow x \in (A \cap B) \cap C.$$

Thus $A \cap (B \cap C) = (A \cap B) \cap C$.

Theorem 2.2.18

Let A be a set. Then

(a) $A \cap \phi = \phi$,

(b) $A \cap U = A$, where U is the universal set.

Proof

(a) Since $\phi \subseteq A$, then $\phi \cap A = \phi$ (by Theorem 2.2.16).

(b) Since $A \subseteq U$, then $A \cap U = A$ (by Theorem 2.2.16).

Theorem 2.2.19 (Distributive law)

Let $A, B,$ and C be sets. Then

(a) $A \cap (B \cup C) = (A \cap B) \cup (A \cap C)$,

(b) $A \cup (B \cap C) = (A \cup B) \cap (A \cup C)$.

Proof

(a) Let $x \in A \cap (B \cup C)$. Then

$$x \in A \cap (B \cup C) \Leftrightarrow x \in A \wedge x \in (B \cup C)$$
$$\Leftrightarrow x \in A \wedge (x \in B \vee x \in C)$$
$$\Leftrightarrow (x \in A \wedge x \in B) \vee (x \in A \wedge x \in C)$$
$$\Leftrightarrow x \in (A \cap B) \vee x \in (A \cap C)$$
$$\Leftrightarrow x \in (A \cap B) \cup (A \cap C).$$

Therefore $A \cap (B \cup C) = (A \cap B) \cup (A \cap C)$.

(b) The proof is left as Exercise 4.

Theorem 2.2.20

Let A and B be sets. Then

(a) $A \subseteq B \Leftrightarrow P(A) \subseteq P(B)$,

(b) $P(A) \cap P(B) = P(A \cap B)$,

(c) $P(A) \cup P(B) \supseteq P(A \cup B)$.

Proof

(a) Suppose $A \subseteq B$ and let $X \in P(A)$. Then

$$X \in P(A) \Rightarrow X \subseteq A$$
$$\Rightarrow X \subseteq B \text{ (since } A \subseteq B)$$
$$\Rightarrow X \in P(B).$$

Thus $P(A) \subseteq P(B)$.
Conversely, suppose that $P(A) \subseteq P(B)$ and let $x \in A$.
Then

$$x \in A \Rightarrow \{x\} \subseteq A$$
$$\Rightarrow \{x\} \in P(A)$$
$$\Rightarrow \{x\} \in P(B) \text{ (since } P(A) \subseteq P(B))$$
$$\Rightarrow x \in B.$$

Thus $A \subseteq B$.

(b) Let $X \in P(A) \cap P(B)$. Then

$$X \in P(A) \cap P(B) \Rightarrow X \in P(A) \wedge X \in P(B)$$
$$\Rightarrow X \subseteq A \wedge X \subseteq B$$
$$\Rightarrow X \subseteq A \cap B$$
$$\Rightarrow X \in P(A \cap B).$$

Thus

$$P(A) \cap P(B) \subseteq P(A \cap B). \tag{2.3}$$

Now, let $X \in P(A \cap B)$. Then

$$X \in P(A \cap B) \Rightarrow X \subseteq A \cap B$$
$$\Rightarrow X \subseteq A \wedge X \subseteq B$$
$$\Rightarrow X \in P(A) \wedge X \in P(B)$$
$$\Rightarrow X \in P(A) \cap P(B).$$

Thus

$$P(A \cap B) \subseteq P(A) \cap P(B). \tag{2.4}$$

From (2.3) and (2.4), we conclude that $P(A) \cap P(B) = P(A \cap B)$.

(c) The proof is left as Exercise 5.

Exercises

1. Complete the proof of Theorem 2.1.6.

2. Complete the proof of Theorem 2.2.7.

3. Complete the proof of Theorem 2.2.17.

4. Complete the proof of Theorem 2.2.19.

5. Complete the proof of Theorem 2.2.20.

6. Let $X \cup Y = X$ for every set X. Show that $Y = \phi$.

7. Describe the set

$$\{x \in \mathbb{R} : x^2 > 2\} \cap \{x \in \mathbb{R} : |x - 2| < |x + 3|\}.$$

8. Let A and B be sets. Prove that

$$\phi \subseteq A \cap B \subseteq A \cup B.$$

9. Consider the following sets

$$A = \{x \in \mathbb{Z}^+ : x = 2y, y \in \mathbb{Z}\},$$
$$B = \{x \in \mathbb{Z}^+ : x = 2y + 1, y \in \mathbb{Z}\},$$
$$C = \{x \in \mathbb{Z}^+ : x = 3y, y \in \mathbb{Z}\}.$$

(a) Describe $A \cap C$ and $B \cup C$.
(b) Verify the validity of $A \cap (B \cup C) = (A \cap B) \cup (A \cap C)$.

10. Let A, B, and C be sets. Prove that

$$(A \cap B) \cup C = A \cap (B \cup C) \text{ iff } C \subseteq A.$$

11. Let A and B be sets. Prove that

(a) $A \cup (A \cap B) = A$,
(b) $A \cap (A \cup B) = A$.

12. Let A, B, and C be sets. Prove that

$$A \cap C = \phi \Rightarrow A \cap (B \cup C) = A \cap B.$$

13. Let A and B be sets. Prove that

$$A \cup B = \phi \Rightarrow A = \phi \wedge B = \phi.$$

14. Let A, B, and C be sets. When

$$A \cup C = B \cup C?$$

15. Let $\#(A)$ be the number of the elements of A. Prove that

$$\#(A \cup B) = \#(A) + \#(B) - \#(A \cap B).$$

16. Let $C \subseteq A$ and $C \subseteq B$. Prove that

$$C \subseteq A \cap B.$$

17. Let X and Y be sets. Prove that if $A \subseteq X$, $B \subseteq Y$, then

$$A \cap B \subseteq X \cap Y.$$

18. Let A and B be sets. Prove that

(a) $A = B \Leftrightarrow P(A) = P(B)$.
(b) $P(A) \cup P(B) \subseteq P(A \cup B)$.
(c) $A \cap B = \phi \Leftrightarrow P(A) \cap P(B) = \{\phi\}$.

19. Let A and B be sets. Give an example to show that

$$P(A) \cup P(B) \neq P(A \cup B).$$

20. Write the power set, $P(X)$, for each of the following sets:

(a) $X = \{a, \{a\}\}$,
(b) $X = \{1, 2, 3, 4\}$,
(c) $X = \{\phi, \{\phi\}\}$,
(d) $X = \{\phi, a, b, \{a, b\}\}$.

21. True or false?

(a) For every set A, $\phi \subseteq A$.
(b) For every set A, $\{\phi\} \subseteq A$.
(c) $\phi \in \{\phi, \{\phi\}\}$.
(d) $\{\phi\} \in \{\phi, \{\phi\}\}$.
(e) $\{\phi\} \subseteq \{\phi, \{\phi\}\}$.
(f) $\{1, 2, 5\} \subseteq \{1, 2, 5, \{6, 7\}\}$.

2.2.3 COMPLEMENT OF SET

Definition 2.2.21
Let U be the universal set and $A \subseteq U$. The set

$$A^C = \{x : x \in U, x \in A\}$$

is said to be the complement of A.

Example 2.2.22
In the Venn diagram, we have shaded the complement of A; that is the area outside of A.

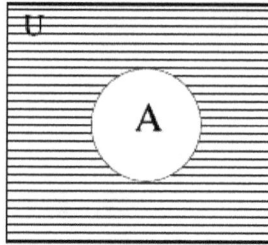

Example 2.2.23
Let the universal set U be $\mathbb{Z}$ and let $A = \{x : x \leq 3\}$. Then

$$A^C = \{4, 5, 6, 7, \ldots\}.$$

Example 2.2.24
Let the universal set U be $\mathbb{N}$ and let

$$A = \{2, 4, 6, 8, \ldots\}.$$

Then

$$A^C = \{1, 3, 5, 7, \ldots\}.$$

Theorem 2.2.25
Let A and B be sets. Then

$$A \subseteq B \Rightarrow B^C \subseteq A^C.$$

Proof
Since $A \subseteq B$, then $x \in A \Rightarrow x \in B$.
Thus $x \notin B \Rightarrow x \notin A$.
That is, $x \in B^C \Rightarrow x \in A^C$.
Hence $B^C \subseteq A^C$.

Theorem 2.2.26
Let A be a set. Then

$$(A^C)^C = A.$$

Proof
It is clear that

$$x \in (A^C)^C \Leftrightarrow x \notin A^C$$
$$\Leftrightarrow x \in A.$$

Thus $(A^C)^C = A$.

Definition 2.2.27
Let A and B be sets. The difference of A and B is defined by

$$A - B = \{x : x \in A \text{ and } x \notin B\}.$$

Remark 2.2.28
$$A - B = A \cap B^C.$$

Example 2.2.29
Let $A = \mathbb{N}$ and let $B = \mathbb{N}_o$. Then
$$A - B = \mathbb{N}_e.$$

Example 2.2.30
Let A and B be sets. Then
$$A - B = B^C - A^C.$$

Proof
We have
$$x \in A - B \Leftrightarrow x \in A \wedge x \notin B$$
$$\Leftrightarrow x \notin A^C \wedge x \in B^C$$
$$\Leftrightarrow x \in B^C - A^C.$$

Thus $A - B = B^C - A^C$.

Theorem 2.2.31
Let A be a set. Then

(a) $U^C = \phi$,

(b) $\phi^C = U$,

(c) $A \cap A^C = \phi$,

(d) $A \cup A^C = U$.

Proof

(a) Since U is the universal set, then
$$U^C = \{x \in U \wedge x U\}$$
$$= \phi.$$

(b) See Exercise 1.

(c)
$$A \cap A^C = \{x \in A \wedge x \in A^C\}$$
$$= \{x \in A \wedge x \notin A\}$$
$$= \phi.$$

(d)

$$A \cup A^C = \{x \in A \vee x \in A^C\}$$
$$= \{x \in A \vee x \in U - A\}$$
$$= U.$$

Theorem 2.2.32 (De Morgan's laws)
Let A and B be sets. Then

(a) $(A \cup B)^C = A^C \cap B^C$,

(b) $(A \cap B)^C = A^C \cup B^C$.

Proof

(a) We have

$$x \in (A \cup B)^C \Leftrightarrow x \notin A \cup B$$
$$\Leftrightarrow x \notin A \wedge x \notin B$$
$$\Leftrightarrow x \in A^C \wedge x \in B^C$$
$$\Leftrightarrow x \in A^C \cap B^C.$$

Thus $(A \cup B)^C = A^C \cap B^C$.

(b) See Exercise 2.

2.2.4   DIFFERENCE AND SYMMETRIC DIFFERENCE OF SETS

Definition 2.2.33
Let A and B be sets. The symmetric difference of A and B is defined by

$$A \Delta B = (A - B) \cup (B - A).$$

Example 2.2.34
In the Venn diagram, we have shaded $A \Delta B$.

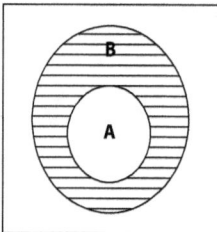
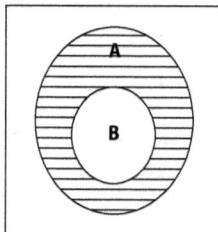
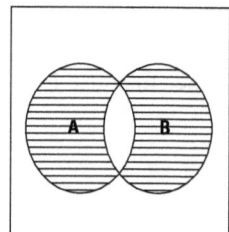

$A \Delta B$ is shaded $A \Delta B$ is shaded $A \Delta B$ is shaded

Theorem 2.2.35
Let A and B be sets. Then

1. $A \Delta \phi = A$,

2. $A \Delta B = \phi \Leftrightarrow A = B$.

Proof

(a)
$$A \Delta \phi = (A - \phi) \cup (\phi - A)$$
$$= A \cup \phi$$
$$= A.$$

(b) Suppose $A \Delta B = \phi$.
Then
$$A \Delta B = \phi \Leftrightarrow (A - B) \cup (B - A) = \phi$$
$$\Leftrightarrow A - B = \phi \wedge B - A = \phi$$
$$\Leftrightarrow A = B.$$

Remark 2.2.36
By using the laws of sets algebra, we can prove the properties of the sets without using the definitions of $\subseteq, \cap,$ and $\cup$.

Example 2.2.37
Prove that $A \cup (A \cap B) = A$.

Proof
$A \cap B \subseteq A$ (by Theorem 2.2.16). Thus
$$A \cup (A \cap B) = A.$$

Example 2.2.38
Prove that $A \cap (A \cup B) = A$.

Proof
We have always $A \subseteq A \cup B$. Thus
$$A \cap (A \cup B) = A \text{ (by Theorem 2.2.16).}$$

Example 2.2.39
Prove that $A \cap (A^C \cup B) = A \cap B$.

Proof
$$A \cap (A^C \cup B) = (A \cap A^C) \cup (A \cap B)$$
$$= \phi \cup (A \cap B)$$
$$= A \cap B.$$

Example 2.2.40

Prove that $A \cup (A \cup B^C)^C = A \cup B$.

Proof

$$
\begin{aligned}
A \cup (A \cup B^C)^C &= A \cup (A^C \cap (B^C)^C) \\
&= A \cup (A^C \cap B) \\
&= (A \cup A^C) \cap (A \cup B) \\
&= U \cap (A \cup B) \\
&= A \cup B.
\end{aligned}
$$

Exercises

1. Complete the proof of Theorem 2.2.31 (b).

2. Complete the proof of Theorem 2.2.32 (b).

3. Let A, B be the subsets of U. Prove that

   (a) $A \subseteq B \Leftrightarrow A \cap B^C = \phi$,

   (b) $A \subseteq B \Leftrightarrow A^C \cup B = U$,

   (c) $A \subseteq B \Leftrightarrow (A \cap B^C) \subset A^C$,

   (d) $A \subseteq B \Leftrightarrow (A \cap B^C) \subset B$.

4. Prove that

   (a) $A \Delta \phi = A$,

   (b) $A \Delta B = B \Delta A$,

   (c) $A \Delta (B \Delta C) = (A \Delta B) \Delta C$,

   (d) $A \cap (B \Delta C) = (A \cap B) \Delta (A \cap C)$,

   (e) $A \Delta A = \phi$,

   (f) $(A \Delta C = B \Delta C) \Rightarrow A = B$.

5. Prove that the equation

   $$(A \cap X) \cup (B \cap X^C) = \phi$$

   has a solution iff $B \subseteq A^C$, and any set X, verifying the relation $B \subseteq X \subseteq A^C$, is a solution of the equation.

6. Prove that the equation

   $$(A \cap X) \cup (B \cap X^C) = (C \cap X) \cup (D \cap X^C)$$

   has a solution iff $B \Delta D \subseteq (A \Delta C^C)$, and find all the solutions.

7. Prove that

   (a) $A \cap (B - C) = (A \cap B) - C$,
   (b) $(A \cup B) - C = (A - C) \cup (B - C)$,
   (c) $A - (B \cup C) = (A - B) \cap (A - C)$,
   (d) $A - (B \cap C) = (A - B) \cup (A - C)$.

8. Prove that

   (a) $A \cup C = B \cup C$ iff $A \Delta B \subseteq C$,
   (b) $(A \cup C) \Delta (B \cup C) = (A \Delta B) - C$.

9. Prove that if $A \subseteq B$ and $C = B - A$, then $A = B - C$.

10. Prove that (Generalized of De Morgan's laws)

   (a) $(A_1 \cup A_2 \cup \cdots \cup A_n)^C = A_1^C \cap A_2^C \cap \cdots \cap A_n^C$,
   (b) $(A_1 \cap A_2 \cap \cdots \cap A_n)^C = A_1^C \cup A_2^C \cup \cdots \cup A_C^n$.

2.3 INDEXED FAMILIES OF SETS

Definition 2.3.1
Let I be a non-empty set. Suppose for each $i \in I$ there is a corresponding set A_i. Then the family of sets

$$A = \{A_i : i \in I\}$$

is an indexed family of sets. Each $i \in I$ is called an index, and I is called an indexing set.

Example 2.3.2
Let $I = \mathbb{N}$. For all $k \in \mathbb{N}$, let

$$A_k = \{1, 2, 3, \ldots, k\}.$$

Then $A_1 = \{1\}$, $A_2 = \{1, 2\}$, $A_3 = \{1, 2, 3\}$, and so forth.
The set with index 12 is

$$A_{12} = \{1, 2, 3, 4, \ldots, 12\}.$$

Example 2.3.3
Let $I = \{1, 2, 3, 4, 5\}$. Consider the sets $A_1 = \{1, 10\}$, $A_2 = \{2, 4, 6, 8, 10\}$, $A_3 = \{3, 6, 9\}$, $A_4 = \{4, 8\}$, $A_5 = \{5, 6, 10\}$. Then $A = \{A_1, A_2, A_3, A_4, A_5\}$ is an indexed family of sets. Furthermore, such an indexed family of sets is denoted by

$$\{A_i\}_{i \in I}.$$

Definition 2.3.4 (Generalized union and intersection)
Let $\{A_i\}_{i \in I}$ be an indexed family of sets. Then

$$\bigcup A_i = \{x : \exists\, i \in I, x \in A_i\}, i \in I,$$
$$\bigcap A_i = \{x : \forall\, i \in I, x \in A_i\}, i \in I.$$

Example 2.3.5
Let $A_1 = \{1, 10\}$, $A_2 = \{2, 4, 6, 10\}$, $A_3 = \{3, 6, 9\}$, $A_4 = \{4, 8\}$, $A_5 = \{5, 6, 10\}$, and $I = \{2, 3, 5\}$. Then

$$\bigcup_{i \in I} A_i = \{2, 3, 4, 5, 6, 9, 10\}, \quad \text{and} \quad \bigcap_{i \in I} A_i = \{6\}.$$

Example 2.3.6
Let $B_n = [0, 1/n]$, where $n \in N$. Then

$$\bigcup_{i \in \mathbb{N}} B_i = [0, 1], \quad \text{and} \quad \bigcap_{i \in \mathbb{N}} B_i = \{0\}.$$

Theorem 2.3.7
Let $\{A_i\}_{i \in I}$ be an indexed family of sets.

(a) If $A_i \subseteq B, \forall\ i \in I$, then $\bigcup_{i \in I} A_i \subseteq B$.

(b) If $B \subseteq A_i, \forall\ i \in I$, then $B \subseteq \bigcup_{i \in I} A_i$.

Proof

(a) Suppose $A_i \subseteq B, \forall\ i \in I$ and let $x \in \bigcup_{i \in I} A_i$.

Then $\exists\, j \in I, x \in A_j$.
Since $A_j \subseteq B$, then $x \in B$.
Thus $\bigcup_{i \in I} A_i \subseteq B$.

(b) Exercise.

Theorem 2.3.8 (Generalized De Morgan's laws) Let $\{A_i\}_{i \in I}$ be an indexed family of sets. Then

(a) $\left(\bigcup_{i \in I} A_i\right)^C = \bigcap_{i \in I} A_i^C,$

(b) $\left(\bigcap_{i \in I} A_i\right)^C = \bigcup_{i \in I} A_i^C.$

Proof

(a) Let $x \in (\bigcup_{i \in I} A_i)^C$.

Then $x \notin \bigcup_{i \in I} A_i^C \Rightarrow x \notin A_i, \forall \; i \in I$

$\Rightarrow x \in A_i, \forall \; i \in I$

$\Rightarrow x \in \bigcap_{i \in I} A_i^C.$

Thus

$$\left(\bigcup_{i \in I} A_i\right)^C \subseteq \bigcap_{i \in I} A_i^C. \tag{2.5}$$

Conversely, suppose that $y \in \bigcap_{i \in I} A_i^C$.

Then $y \in A_i^C, \forall i \in I \Rightarrow y \notin A_i, \forall \; i \in I$

$\Rightarrow y \notin \bigcup_{i \in I} A_i$

$\Rightarrow y \in (\bigcup_{i \in I} A_i)^C.$

Thus

$$\bigcap_{i \in I} A_i^C \subseteq \left(\bigcup_{i \in I} A_i\right)^C. \tag{2.6}$$

It follows from Equations (2.5) and (2.6) that

$$\left(\bigcup_{i \in I} A_i\right)^C = \bigcap_{i \in I} A_i^C.$$

(b) Exercise.

Theorem 2.3.9 (Generalized distribution law) Let $\{A_i\}_{i \in I}$ and $\{B_j\}_{j \in J}$ be indexed families of sets. Then

(a) $\left(\bigcup_{i \in I} A_i\right) \cap \left(\bigcup_{j \in J} B_j\right) = \bigcup_{(i,j) \in I \times J} (A_i \cap B_j),$

(b) $\left(\bigcap_{i \in I} A_i\right) \cup \left(\bigcap_{j \in J} B_j\right) = \bigcap_{(i,j) \in I \times J} (A_i \cup B_j).$

Proof

(a) Suppose $x \in \left(\bigcup_{i \in I} A_i\right) \cap \left(\bigcup_{j \in J} B_j\right)$

$\Rightarrow (\exists \, h \in I, x \in A_h) \wedge (\exists \, k \in J, x \in B_k)$

$\Rightarrow \exists (h, k) \in I \times J, x \in A_h \cap B_k$

$\Rightarrow \exists x \in \bigcup_{(i,j) \in I \times J} (A_i \cap B_j).$

Thus

$$\left(\bigcup_{i \in I} A_i\right) \cap \left(\bigcup_{j \in J} B_j\right) = \bigcup_{(i,j) \in I \times J} (A_i \cap B_j). \tag{2.7}$$

Conversely, let $y \in \bigcup_{(i,j) \in I \times J} (A_i \cap B_j)$.

Then, there exists $(s,t) \in I \times J$, such that $y \in A_s \cap B_t$, that is $(\exists s \in I \wedge t \in J)$ such that $(y \in A_s, y \in B_t)$. Thus

$$(\exists s \in I, y \in A_s) \wedge (\exists t \in J, y \in B_t).$$

Therefore

$$y \in \bigcup_{i \in I} A_i \wedge y \in \bigcup_{j \in J} B_j$$

$$\Rightarrow y \in \left(\bigcup_{i \in I} A_i\right) \cap \left(\bigcup_{j \in J} B_j\right)$$

$$\Rightarrow \bigcup_{(i,j) \in I \times J} (A_i \cap B_j) \subseteq \left(\bigcup_{i \in I} A_i\right) \cap \left(\bigcup_{j \in J} B_j\right). \tag{2.8}$$

From Equations (2.7) and (2.8), we conclude that

$$\left(\bigcup_{i \in I} A_i\right) \cap \left(\bigcup_{j \in J} B_j\right) = \bigcup_{(i,j) \in I \times J} (A_i \cap B_j).$$

Exercises

1. Let $\{A_i\}_{i \in I}$ and $\{B_j\}_{j \in J}$ be indexed families of sets. Prove that

   (a) $\left(\bigcup_{i \in I} A_i\right) - \left(\bigcup_{j \in J} B_j\right) = \bigcup_{i \in I} [\bigcap_{j \in J} (A_i - B_j)]$,

   (b) $\left(\bigcap_{i \in I} A_i\right) - \left(\bigcap_{j \in J} B_j\right) = \bigcap_{i \in I} [\bigcup_{j \in J} (A_i - B_j)]$.

2. The indexed family of sets $\{B_i\}_{i \in I}$ is said to be a covering of a set A if $A \subseteq \bigcup_{j \in J} B_j$. Prove that if $\{B_i\}_{i \in I}$ and $\{C_j\}_{j \in J}$ are distinct coverings of a set A, then

$$\{B_i \cap C_j\}_{(i,j) \in I \times J}$$

   is also a covering of A.

3. Consider

$$A_\alpha = 1, \frac{1}{2}, \frac{1}{3}, \ldots, \frac{1}{\alpha}, \ \forall \, \alpha \in \mathbb{Z}^+$$

   For $X = \{x \in \mathbb{R} : 0 \leq x \leq 1\}$ to find

   (a) $\cap \{A_\alpha : \alpha \in \mathbb{Z}^+\}$,

   (b) $\cup \{X - A_\alpha : \alpha \in \mathbb{Z}^+\}$,

   (c) $\cup \{A_\alpha : \alpha \in \mathbb{Z}^+\}$,

   (d) $\cup \{X - A_\alpha : \alpha \in \mathbb{Z}^+\}$.

4. Let $\{A_\alpha : \alpha \in \mathbb{Z}^+\}$ be an indexed family of subsets of a set X, such that

$$\cap\{A_\alpha : \alpha \in \mathbb{Z}^+\} = \phi.$$

Prove that

$$\cup\{X - A_\alpha : \alpha \in \mathbb{Z}^+\} = X.$$

5. Let $A = \{A_i : i \in I\}$ be an arbitrary family of sets and let B be a set. Prove that

(a) $B \cap \left(\bigcup_{i \in I} A_i\right) = \bigcup_{i \in I}(B \cap A_i),$

(b) $B \cup \left(\bigcap_{i \in I} A_i\right) = \bigcap_{i \in I}(B \cap A_i).$

6. Let $A = \{A_i : i \in I\}$ be a family of sets and $J \subseteq I$. Prove that

(a) $\bigcup_{i \in J} A_i \subseteq \bigcup_{i \in I} A_i,$

(b) $\bigcap_{i \in I} A_i \subseteq \bigcap_{i \in J} A_i.$

7. Let A be a family of sets and suppose that $A \subseteq B, \forall\ A \in A$. Prove that

$$\bigcup_{A \in A} A \subseteq B.$$

8. Let A be a family of sets and suppose that $B \subseteq A, \forall\ A \in A$. Prove that

$$B \subseteq \bigcap_{A \in A} A.$$

9. Find the union and intersection of each of the following families or indexed collections A, where

(a) $A = \{\{1,2,3,4,5\}, \{2,3,4,5,6\}, \{3,4,5,6,7\}, \{4,5,6,7,8\}\},$

(b) $A = \{\{1,3,5\}, \{2,4,6\}, \{7,9,11,13\}, \{8,10,12\}\},$

(c) $A = \{A_n : n \in \mathbb{N}\}, A_n = \{1,2,3,\ldots,n\}, \forall\ n \in \mathbb{N},$

(d) $A = \{A_n : n \in \mathbb{N}\}, A_n = N - \{1,2,3,\ldots,n\}, \forall\ n \in \mathbb{N},$

(e) A is the collection of all sets of integers that contain 10,

(f) $A = \{A_n : n \in \{1,2,3,\ldots,10\}\}, A_1 = \{1\}, A_2 = \{2,3\}, A_3 = \{3,4,5\}, \ldots,$
$A_{10} = \{10,11,\ldots,19\},$

(g) $A = \{A_n : n \in \mathbb{N}\}, A_n = (0, \frac{1}{n}), \forall\ n \in N,$

(h) $A = \{A_r : r \in \mathbb{R}^+\}, \mathbb{R}^+ = (0,\infty), A_r = [-\pi, r)$ for $r \in \mathbb{R}^+,$

(i) $A = \{A_r : r \in \mathbb{R}\}, A_r = [|r|, 2|r| + 1], \forall\ r \in \mathbb{R},$

(j) $A = \{A_n : n \geq 3\}, A_n = [\frac{1}{n}, 2 + \frac{1}{2}], \forall\ n \geq 3,$

(k) $A = \{C_n : n \in \mathbb{Z}\}, C_n = [n, n+1], \forall\ n \in \mathbb{Z}.$

3 Relations

If x and y are integers, we might say that x is related to y when x is less than y. In this chapter, we will study the idea of "is related to" by making the notion of a relation precise.

3.1 ORDERED PAIRS AND CARTESIAN PRODUCT

The study of relations begins with the concept of an ordered pair.

Definition 3.1.1
An ordered pair consists of two elements, say a and b, in which one of them, say a, is designated as the first element and the other as the second element. An ordered pair is denoted by (a,b).

Example 3.1.2
The ordered pairs (5, 6) and (6, 5) are different,
that is (5, 6) $\neq$ (6, 5).

Example 3.1.3
The set $\{5,6\}$ is not an ordered pair since the elements 5 and 6 are not designated.

Remark 3.1.4
An ordered pair (a,b) can be defined rigorously by

$$(a,b) \equiv \{a,\{a,b\}\},$$

from this definition, the fundamental property of ordered pairs can be proven:

$$(a,b) = (c,d) \Leftrightarrow a = c \wedge b = d.$$

Definition 3.1.5
Let A and B be sets. The Cartesian product of A and B consists of all ordered pairs (a,b), where $a \in A$ and $b \in B$. It is denoted by

$$A \times B,$$

which reads "A cross B". More concisely,

$$A \times B = \{(a,b) : a \in A, b \in B\}.$$

Example 3.1.6
Let $A = \{a,b\}$ and $B = \{1,2,3\}$. Then

$$A \times B = \{(a,1),(a,2),(a,3),(b,1),(b,2),(b,3)\}.$$

DOI: 10.1201/9780429022838-3

Example 3.1.7

Let $C = \{x, y\}$. Then

$$C \times C = \{(x, x), (x, y), (y, x), (y, y)\}.$$

Example 3.1.8

Let $A = \mathbb{R}$ the set of real numbers. Then

$$\mathbb{R} \times \mathbb{R} = \{(x, y) : x, y \in \mathbb{R}\}.$$

The elements of $\mathbb{R} \times \mathbb{R}$ are the points of the xy-plane (Cartesian plane).

Remark 3.1.9

1. If the set A has n elements and the set B has m elements, then $A \times B$ has nm elements.

2. The Cartesian product of two sets is not commutative, that is $A \times B \neq B \times A$.

3. If A or B is empty set, then $A \times B = \phi$.

4. If either A or B is infinite and the other is not empty, then $A \times B$ is infinite.

The concept of the Cartesian product can be extended to more than two sets in a natural way.

The Cartesian product of sets A, B, and C, denoted by

$$A \times B \times C,$$

consists of all ordered triples (a, b, c), where $a \in A$, $b \in B$, and $c \in C$; that is, $A \times B \times C = \{(a, b, c) : a \in A, b \in B, \text{ and } c \in C\}$.

Analogously, the Cartesian product of n sets

$$A_1, A_2, \ldots, A_n$$

denoted by

$$A_1 \times A_2 \times \cdots \times A_n,$$

consists of all ordered n-tuples $(a_1, a_2, \ldots, a_n)$, where

$$a_1 \in A_1, a_2 \in A_2, \ldots, a_n \in A_n,$$

that is, $A_1 \times A_2 \times \cdots \times A_n = \{(a_1, a_2, \ldots, a_n) : a_i \in A_i \ \forall \ i = 1, 2, \ldots, n\}$.

Example 3.1.10

Let $A = \{x, y\}$, $B = \{1, 2, 3\}$, and $C = \{a, b\}$. Then

$$\begin{aligned}A \times B \times C = \{&(x, 1, a), (x, 1, b), (x, 2, a), (x, 2, b), (x, 3, a), (x, 3, b), \\ &(y, 1, a), (y, 1, b), (y, 2, a), (y, 2, b), (y, 3, a), (y, 3, b)\}.\end{aligned}$$

Example 3.1.11

If $A_i = \mathbb{R}$, $1 \leq i \leq n$. Then

$$A_1 \times A_2 \times \cdots \times A_n = \{(a_1, a_2, \ldots, a_n) : a_i \in \mathbb{R}, 1 \leq i \leq n\}$$

and is denoted by $\mathbb{R}^n$.

Theorem 3.1.12

Let A and B be nonempty sets. Then

$$A \times B = B \times A \text{ iff } A = B$$

Proof

Suppose $A \times B = B \times A$.

Let $a \in A$, then

$\forall \ b \in B$, $(a, b) \in A \times B$

$$\Rightarrow (a, b) \in B \times A$$
$$\Rightarrow a \in B \wedge b \in A.$$

Hence $a \in A \Rightarrow a \in B$, and

therefore, $A \subseteq B$

Similarly we prove that $B \subseteq A$.

Thus $A = B$.

Conversely, suppose that $A = B$.

It is clear that $A \times A = A \times A$.

Thus $A \times B = B \times A$.

Theorem 3.1.13

Let A, B, and C be sets. Then

   (a) $A \times (B \cap C) = (A \times B) \cap (A \times C)$,

   (b) $A \times (B \cup C) = (A \times B) \cup (A \times C)$,

   (c) $(A \times B) \cap (C \times D) = (A \cap C) \times (B \cap D)$.

Proof

   (a) Suppose $(x, y) \in A \times (B \cap C)$. Then

$$(x, y) \in A \times (B \cap C) \Leftrightarrow x \in A \wedge y \in (B \cap C)$$
$$\Leftrightarrow x \in A \wedge [y \in B \wedge y \in C]$$
$$\Leftrightarrow (x \in A \wedge y \in B) \wedge (x \in A \wedge y \in C)$$
$$\Leftrightarrow (x, y) \in A \times B \wedge (x, y) \in A \times C$$
$$\Leftrightarrow (x, y) \in (A \times B) \cap (A \times C).$$

   (b) See Exercise 1.

(c) It is obvious that

$$
\begin{aligned}
(x,y) \in (A \times B) \cap (C \times D) &\Rightarrow (x,y) \in A \times B \wedge (x,y) \in C \times D \\
&\Rightarrow (x \in A \wedge y \in B) \wedge (x \in C \wedge y \in D) \\
&\Rightarrow (x \in A \wedge x \in C) \wedge (y \in B \wedge y \in D) \\
&\Rightarrow x \in (A \cap C) \wedge y \in (B \cap D) \\
&\Rightarrow (x,y) \in (A \cap C) \times (B \cap D).
\end{aligned}
$$

Thus

$$(A \times B) \cap (C \times D) \subseteq (A \cap C) \times (B \cap D) \tag{3.1}$$

Conversely,

$$
\begin{aligned}
(x,y) \in (A \cap C) \times (B \cap D) &\Rightarrow x \in (A \cap C) \wedge y \in (B \cap D) \\
&\Rightarrow (x \in A \wedge x \in C) \wedge (y \in B \wedge y \in D) \\
&\Rightarrow (x \in A \wedge y \in B) \wedge (x \in C \wedge y \in D) \\
&\Rightarrow (x,y) \in A \times B \wedge (x,y) \in C \times D \\
&\Rightarrow (x,y) \in (A \times B) \cap (C \times D).
\end{aligned}
$$

Thus

$$(A \cap C) \times (B \cap D) \subseteq (A \times B) \cap (C \times D) \tag{3.2}$$

From Equations (3.1) and (3.2) we conclude that

$$(A \times B) \cap (C \times D) = (A \cap C) \times (B \cap D).$$

EXERCISES

1. Prove the part (b) of Theorem 3.1.13.

2. If $A, B,$ and C are sets, prove that

   (a) $A \times \phi = \phi$,
   (b) $(A \times B) \cup (C \times D) \subseteq (A \cup C) \times (B \cup D)$.

3. Find a and b if

   $$(a+b, 3a+5b) = (a-b, 2a-7b).$$

4. Prove that if $(x,y,z) = (u,v,w)$, then

   $$x = u \wedge y = v \wedge z = w.$$

5. Let $A = \{1,3,5,7\}, B = \{-2,-9,6\}, C = \{x,y,z\}$. Find

   (a) $(A \cup B) \times C$,
   (b) $(A \times C) \cup (B \times C)$,
   (c) $(A \cup B) \times (B \cup C)$.

6. For arbitrary sets A, B, and C to prove that

(a) $(A \times A) \cap (B \times C) = (A \cap B) \times (A \cap C)$,
(b) $(A \times B) - (C \times C) = [(A - C) \times B] \cup [A \times (B - C)]$,
(c) $(A \times A) - (B \times C) = [(A - B) \times A] \cup [A \times (A - C)]$.

7. For arbitrary sets A, B, C, and D to prove that

(a) $A \times (B - D) = (A \times B) - (A \times D)$,
(b) $(A \times B) \cap (C \times V) = (A \times D) \cap (C \times B)$.

8. Let A, B, and C are sets, such that $A \neq \phi$, $B \neq \phi$, and

$$(A \times B) \cup (B \times A) = C \times C.$$

Prove that $A = B = C$.

9. Let A and B be sets. Prove that

$$A \cap B = \phi \text{ iff } (A \times C) \cap (B \times C) = \phi,$$

for any nonempty set C.

10. Let A, B, and C are sets, such that $A \subseteq B$. Prove that

$$A \times C \subseteq B \times C.$$

11. Prove that

$$\{1, 2, 3, \ldots, n\} \times A = (\{1\} \times A) \cup (\{2\} \times A) \cup \cdots \cup (\{n\} \times A)$$

for an arbitrary set A.

12. Give an example of sets A, B, and C, such that

$$A \cup (B \times C) \neq (A \cup B) \times (A \cup C).$$

13. Prove that if $A \times A = B \times B$, then $A = B$.

14. Prove that if $X \times Y = X \times Z$ and $X \neq \phi$, then $Y = Z$.

15. Let A and C be nonempty sets. Prove that

$$A \subseteq B \wedge C \subseteq D \text{ iff } A \times C \subseteq B \times D.$$

16. Let A, B, C, and D be nonempty sets. Prove that

$$A \times B = C \times D \Leftrightarrow A = C \wedge B = D.$$

17. Let $A = \{a, b\}$, $B = \{2, 3\}$, and $C = \{3, 4\}$. Find

(a) $A \times (b \cup C)$,
(b) $(A \times B) \cup (A \times C)$,
(c) $A \times (B \cap C)$,
(d) $(A \times B) \cap (A \times C)$.

18. Suppose that the ordered pairs $(x + y, 1)$ and $(3, x - y)$ are equal. Find x and y.

3.2   RELATIONS ON SETS

Definition 3.2.1

Let A and B be sets. Then any subset of $A \times B$ is called a relation from A to B.
If R is a relation from A to B, then

$$R \subseteq A \times B.$$

Subsets of $A \times A$ are called relations on A.
If R is a relation from A to B and $(a,b) \in R$, we write aRb and we read "a is related to b". $a \not{R} b$ means that $(a,b) \notin R$.

Example 3.2.2

Let $A = \{1,5\}$, $B = \{2,4,6\}$, then

$$R = \{(1,2),(1,6),(5,6)\} \text{ is a relation from } A \text{ to } B.$$

We have $1R2, 5R6$. Also the following sets
$S = \{(2,1),(2,5),(6,1),(6,5)\}$
$T = \{(4,1)\}$ are relations from B to A.

Example 3.2.3

Let $X = \{a,b,c,3\}$, $Y = \{1,2,5\}$, then

$$R = \{(a,1),(b,1),(c,2),(3,2),(3,5)\}$$

is a relation from X to Y. We have $aR1, 3R2$, and $3R5$. However, $3 \not{R} 1$ and $b \not{R} 2$. The sets

$$S = \{(1,a),(2,3),(5,a)\}$$
$$T = \{(2,c)\}$$

are relations from Y to X.

Example 3.2.4

Let $R = \{(x,y) \in \mathbb{R} \times \mathbb{R} : x \leq y\}$. Then
$(1,1) \in R$ and $(3,7) \in R$. But $(3,2) \notin R$, and $(10,6) \notin R$.

Example 3.2.5

Let X be a set. Then the set

$$R = \{(A,B) \in P(X) \times P(X) : A \subseteq B\}$$

is a relation on the set $P(X)$.

Example 3.2.6

Let A be the set of the lines in the xy-plane. The set

$$R = \{(a,b) \in A \times A : a // b\}$$

is a relation on A.

Example 3.2.7
Consider the relation R on the set $\mathbb{N} \times \mathbb{N}$ given by

$$R = \{((a,b),(c,d)) \in (\mathbb{N} \times \mathbb{N}) \times (\mathbb{N} \times \mathbb{N}) : a+b=c+d\}.$$

Then $(1,5)R(2,4)$ but $(6,7)\not\!R\,(4,10)$.

Remark 3.2.8
Since the relation is a set, then we can define union, intersection, and difference of two relations.

For example, if R and S are relations from A to B, that is

$$R \subseteq A \times B \text{ and } S \subseteq A \times B, \text{ then } R \cup S \subseteq A \times B, R \cap S \subseteq A \times B,$$

and $R - S \subseteq A \times B$. Also $R \cup S$, $R \cap S$, and $R - S$ are relations from A to B. In symbols,

$$R \cup S = \{(x,y) \in A \times B : (x,y) \in R \vee (x,y) \in S\},$$
$$R \cap S = \{(x,y) \in A \times B : (x,y) \in R \wedge (x,y) \in S\},$$
$$R - S = \{(x,y) \in A \times B : (x,y) \in R \wedge (x,y) \notin S\}.$$

Example 3.2.9
Let

$$R = \{(x,y) \in \mathbb{R} \times \mathbb{R} : x+y=5\},$$
$$S = \{(x,y) \in \mathbb{R} \times \mathbb{R} : 2x-y=4\}.$$

Then

$$R \cap S = \{(x,y) \in \mathbb{R} \times \mathbb{R} : x+y=5 \wedge 2x-y=4\}$$
$$= \{(3,2)\}.$$

Definition 3.2.10
Let R be a relation from A to B, the domain of R, denoted by $\text{Dom}(R)$, is

$$\text{Dom}(R) = \{a \in A : \exists\, b \in B, \text{ such that } (a,b) \in R\}.$$

The range of R, denoted by $\text{Rng}(R)$, is

$$\text{Rng}(R) = \{b \in B : \exists\, a \in A, \text{ such that } (a,b) \in R\}.$$

Thus, the domain of R is the set of all first coordinates of ordered pairs in R, and the range of R is the set of all second coordinates. By definition, $\text{Dom}(R) \subseteq A$ and $\text{Rng}(R) \subseteq B$.

Example 3.2.11

Let $A = \{1,2,3,4\}$, $B = \{x,y,z\}$, and

$$R = \{(1,x),(2,x),(2,z),(4,z)\}.$$

Then

$$\text{Dom}(R) = \{1,2,4\},$$

and

$$\text{Rng}(R) = \{x,z\}.$$

Example 3.2.12

Let R be a relation on $\mathbb{R}$ defined by

$$R = \{(x,y) \in \mathbb{R} \times \mathbb{R} : y = x^2\}.$$

Then

$$\begin{aligned}
\text{Dom}(R) &= \{x \in \mathbb{R} : \exists\, y \in \mathbb{R}, \text{ such that } (x,y) \in R\} \\
&= \{x \in \mathbb{R} : \exists\, y \in \mathbb{R}, \text{ such that } y = x^2\}.
\end{aligned}$$

Then $\text{Dom}(R) = \mathbb{R}$ and

$$\begin{aligned}
\text{Rng}(R) &= \{y \in R : \exists\, x \in \mathbb{R}, \text{ such that } (x,y) \in R\} \\
&= \{y \in R : \exists\, x \in \mathbb{R}, \text{ such that } y = x^2\}.
\end{aligned}$$

Then,

$$\text{Rng}(R) = \{y \in R : y \geq 0\}.$$

Definition 3.2.13

Let R be a relation from A to B, then the inverse of R is

$$R^{-1} = \{(y,x) : (x,y) \in R\}.$$

Example 3.2.14

Let $R = \{(1,b),(1,c),(3,c)\}$. Then

$$R^{-1} = \{(b,1),(c,1),(c,3)\}.$$

Example 3.2.15

Let $R = \{(x,y) \in \mathbb{R} \times \mathbb{R} : x \leq y\}$. Then

$$R^{-1} = \{(x,y) \in \mathbb{R} \times \mathbb{R} : x \geq y\}.$$

Example 3.2.16

Let S be a relation on $A = \mathbb{N} \cup \{0\}$, defined by

$$\begin{aligned}
S &= \{(a,b) \in A \times A : a = 0\} \\
&= \{(0,0),(0,1),(0,2),\dots\}.
\end{aligned}$$

Then

$$S^{-1} = \{(x,y) \in A \times A : y = 0\}$$
$$= \{(0,0), (1,0), (2,0), \ldots\}.$$

Example 3.2.17
Let T be a relation on $\mathbb{R}$, defined by

$$T = \{(x,y) \in \mathbb{R} \times \mathbb{R} : y = x^2\}.$$

Then

$$T^{-1} = \{(x,y) \in \mathbb{R} \times \mathbb{R} : x = y^2\}.$$

Theorem 3.2.18
Let R be a relation on A. Then
$$(R^{-1})^{-1} = R.$$

Proof
Let $(x,y) \in (R^{-1})^{-1}$, then

$$(x,y) \in (R^{-1})^{-1} \Rightarrow (y,x) \in R^{-1}$$
$$\Rightarrow (x,y) \in R.$$

Thus

$$(R^{-1})^{-1} \subseteq R \qquad\qquad (3.3)$$

Now, let $(x,y) \in R$. Then

$$(x,y) \in R \Rightarrow (y,x) \in R^{-1}$$
$$\Rightarrow (x,y) \in (R^{-1})^{-1}.$$

Hence

$$R \subseteq (R^{-1})^{-1} \qquad\qquad (3.4)$$

From Equations (3.3) and (3.4), we conclude that

$$(R^{-1})^{-1} = R.$$

Theorem 3.2.19
If R is a relation from A to B, then

(a) $\mathrm{Dom}(R) = \mathrm{Rng}(R^{-1})$,

(b) $\mathrm{Rng}(R) = \mathrm{Dom}(R^{-1})$.

Proof

(a) We have

$$x \in \text{Dom}(R) \Rightarrow \exists y \in B \text{ such that } (x,y) \in R$$
$$\Rightarrow \exists y \in B \text{ such that } (y,x) \in R^{-1}$$
$$\Rightarrow x \in \text{Rng}(R^{-1}).$$

Then

$$\text{Dom}(R) \subseteq \text{Rng}(R^{-1}) \qquad (3.5)$$

Now, let $x \in \text{Rng}(R^{-1})$. Then

$$x \in \text{Rng}(R^{-1}) \Rightarrow \exists y \in B, \text{ such that } (y,x) \in R^{-1}$$
$$\Rightarrow y \in B, \text{ such that } (x,y) \in R \qquad \Rightarrow x \in \text{Dom}(R).$$

Thus,

$$\text{Rng}(R^{-1}) \subseteq \text{Dom}(R) \qquad (3.6)$$

From Equations (3.5) and (3.6), we deduce that

$$\text{Dom}(R) = \text{Rng}(R^{-1}).$$

(b) See Exercise 1.

Definition 3.2.20
Let R be a relation from A to B and let

$$C \subseteq A, D \subseteq B. \text{ Then the set } R \cap (C \times D) \text{ is}$$

called restriction of relation R from C to D.

Remark 3.2.21
If R is a relation on A and $C = D$, then

$$R \cap (C \times C) \text{ is called restriction of } R \text{ on } C,$$

denoted by $R|_C$.

Example 3.2.22
Let $A = \{x \in \mathbb{N} : x \text{ is even}\}$,
    $B = \{x \in \mathbb{N} : x \text{ is odd}\}$,
    $C = \{2,4,6\}$,
and
    $D = \{1,3,5\}$.

If R is a relation from A to B, defined by

$$R = \{(x,y) : x \in A \land y \in B, \text{ such that } y|x\},$$

then the restriction of R from C to D is the set

$$R \cap (C \times D) = \{(2,1),(4,1),(6,1),(6,3)\}.$$

Example 3.2.23
Let

$$A = \{x \in \mathbb{Z} : -16 \leq x \leq 16\},$$
$$B = \{x \in \mathbb{N} : x \leq 16\}.$$

If R is a relation on A defined by

$$R = \{(x,y) \in A \times A : y = x^2 + 1\},$$

then $R|_B = \{(0,1),(1,2),(2,5),(3,10)\}$.

Definition 3.2.24
If R is a relation from A to B and S is a relation from B to C, then the composite of R and S, denoted by $S \circ R$, is defined by

$$S \circ R = \{(x,z) : \exists\, y \in B, \text{ such that } (x,y) \in R \text{ and } (y,z) \in S\}.$$

Since $S \circ R \subseteq A \times C$, then $S \circ R$ is a relation from A to C.

Example 3.2.25 Let $A = \{1,2,3,4\}$, $B = \{x,y,z,w\}$, and $C = \{a,b,c\}$.
Let R be a relation from A to B, defined by

$$R = \{(1,x),(1,y),(2,y),(3,z),(4,w)\},$$

and S a relation from B to C, defined by

$$S = \{(x,a),(y,a),(y,b),(w,c)\}.$$

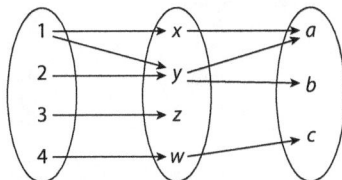

The relations R and S are illustrated in the figure above. We have

$$S \circ R = \{(1,a),(1,b),(2,a),(2,b),(4,c)\}.$$

Example 3.2.26 Let

$$R = \{(x,y) \in \mathbb{R} \times \mathbb{R} : y = x+1\},$$
$$S = \{(x,y) \in \mathbb{R} \times \mathbb{R} : y = x^2\}.$$

Then

$$R \circ S = \{(x,y) : \exists z \in \mathbb{R}, \text{ such that } (x,z) \in S \wedge (z,y) \in R\}$$
$$= \{(x,y) : z = x^2 \wedge y = z+1\}$$
$$= \{(x,y) : y = x^2 + 1\},$$

and

$$S \circ R = \{(x,y) : \exists z \in \mathbb{R}, \text{ such that } (x,z) \in R \wedge (z,y) \in S\}$$
$$= \{(x,y) : z = x+1 \wedge y = z^2\}$$
$$= \{(x,y) : y = (x+1)^2\}.$$

Clearly, $S \circ R \neq R \circ S$.

Theorem 3.2.27 Let R be a relation on A. Then

$$I_A \circ R = R \circ I_A = R,$$

where I_A is the identity relation on A, that is

$$I_A = \{(x,x) : x \in A\}.$$

Proof We have

$$(x,y) \in I_A \circ R \Rightarrow \exists z \in A, \text{ such that } (x,z) \in R \wedge (z,y) \in I_A.$$

Since $(z,y) \in I_A$, then $z = y$. Thus $(x,z) \in R \Rightarrow (x,y) \in R$.
Hence

$$I_A \circ R \subseteq R \tag{3.7}$$

Now, let $(x,y) \in R$. Since
$(x,y) \in R \wedge (y,y) \in I_A$,
then $(x,y) \in I_A \circ R$. Thus

$$R \subseteq I_A \circ R \tag{3.8}$$

From Equations (3.7) and (3.8), we conclude that $I_A \circ R = R$.
Analogously, we prove that $R \circ I_A = R$.

Theorem 3.2.28 Let $R, S,$ and T be relations on a set A. Then

(a) $(T \circ S) \circ R = T \circ (S \circ R),$

(b) $(S \cup T) \circ R = (S \circ R) \cup (T \circ R),$

(c) $(S \cap T) \circ R = (S \circ R) \cup (T \circ R),$

(d) If $R \subseteq S$, then $T \circ R \subseteq T \circ S$ and
$R \circ T \subseteq S \circ T,$

(e) $(S \circ R) \cap T = \phi \Rightarrow (T \circ R^{-1}) \cap S = \phi,$

(f) $(S \circ R)^{-1} = R^{-1} \circ S^{-1}.$

Proof (a) Suppose that $(x, w) \in (T \circ S) \circ R.$
Then $\exists \ y$, such that $(x, y) \in R \wedge (y, w) \in T \circ S$
and $\exists \ z$, such that $(y, z) \in S \wedge (z, w) \in T.$
Since
$$(x, y) \in R \wedge (y, z) \in S,$$
then $(x, z) \in S \circ R.$
As $(x, z) \in S \circ R$ and $(z, w) \in T$, then $(x, w) \in T \circ (S \circ R).$ Thus
$$(T \circ S) \circ R \subseteq T \circ (S \circ R). \tag{3.9}$$

By the same manner, we show that
$$T \circ (S \circ R) \subseteq (T \circ S) \circ R. \tag{3.10}$$

From Equations (3.9) and (3.10), we get
$$T \circ (S \circ R) = (T \circ S) \circ R.$$

(b) See Exercise 2.
(c) See Exercise 2.
(d) See Exercise 2.
(e) Suppose that $(S \circ R) \cap T \neq \phi$. Then $\exists \ (x, y)$ such that $(x, y) \in (S \circ R) \cap T$; i.e.
$(x, y) \in S \circ R \wedge (x, y) \in T.$
Since $(x, y) \in S \circ R$, then $\exists \ z$ such that $(x, z) \in R \wedge (z, y) \in S.$ $(x, z) \in R \Rightarrow (z, x) \in R^{-1}.$
Then $((z, x) \in R^{-1} \wedge (x, y) \in T) \Rightarrow (z, y) \in T \circ R^{-1}.$ As $(z, y) \in S$, then $(z, y) \in (T \circ R^{-1}) \cap S$, i.e., $(T \circ R^{-1}) \cap S \neq \phi.$
Analogously we prove that
$$(T \circ R^{-1}) \cap S \neq \phi \Rightarrow (S \circ R) \cap T \neq \phi.$$

Thus
$$(S \circ R) \cap T = \phi \Leftrightarrow (T \circ R^{-1}) \cap S = \phi.$$

Exercises

1. Complete the proof of Theorem 3.2.19.

2. Complete the proof of Theorem 3.2.28.

3. Let $X = \{a,b,c\}$, and $Y = \{r\}$. Find all the relations from X to Y.

4. Suppose that A has n elements. How many relations are there on A?

5. Let S be a relation from X to Y, T be a relation from Y to Z and let $A \subseteq X$. Define
$$S(A) = \{y : (x,y) \in S,\ x \in A\}.$$
Prove that

(a) $S(A) \subseteq Y$,
(b) $(T \circ S)(A) = T(S(A))$,
(c) $S(A \cup B) = S(A) \cup S(B)$,
(d) $S(A \cap B) = S(A) \cap S(B)$.

6. Let S and T be relations from X to Y. Prove that

(a) $(S \cap T)^{-1} = S^{-1} \cap T^{-1}$,
(b) $(S \cup T)^{-1} = S^{-1} \cup T^{-1}$.

7. Show by an example that $S \circ R \neq R \circ S$, where R and S are relations on a set A.

8. Let G be a relation from X to Y and H be a relation from Y to Z. Prove that

(a) $\mathrm{Dom}(G \circ H) \subseteq \mathrm{Dom}(H)$,
(b) $\mathrm{Rng}(G \circ H) \subseteq \mathrm{Rng}(G)$,
(c) if $\mathrm{Rng}(H) \subseteq \mathrm{Rng}(G)$, then $\mathrm{Dom}(G \circ H) = \mathrm{Dom}(H)$.

9. Let G,H,J, and K be relations on a set A. Prove that

(a) $G \subseteq H \wedge J \subseteq K \Rightarrow G \circ J \subseteq H \circ K$,
(b) $G \subseteq H \Leftrightarrow G^{-1} \subseteq H^{-1}$.

10. Let G and H be relations on a set A. Prove that

(a) $\mathrm{Dom}(G \cup H) = \mathrm{Dom}(G) \cup \mathrm{Dom}(H)$,
(b) $\mathrm{Rng}(G \cup H) = \mathrm{Rng}(G) \cup \mathrm{Rng}(H)$.

11. Let R be a relation on a set A and let $B,C \subseteq A$. Prove that

(a) $R|_{B \cap C} = (R|_B) \cap (R|_C)$,
(b) $R|_{B \cup C} = (R|_B) \cup (R|_C)$.

12. Find the domain and range for the relation T on $\mathbb{R}$ given by xTy iff

(a) $y = 2x + 1$,   (b) $y = \sqrt{x-1}$,
(c) $y \leq x^2$,   (d) $|x| < 2 \wedge y = 3$,
(e) $y = x^2 + 3$,   (f) $y = \frac{1}{x^2}$,
(g) $y \neq x$,   (h) $|x| < 2 \wedge y = 3$.

13. Find the inverse of the following relations:

(a) $R_1 = \{(x,y) \in \mathbb{R} \times \mathbb{R} : y = x\}$,
(b) $R_2 = \{(x,y) \in \mathbb{R} \times \mathbb{R} : y = -5x + 2\}$,
(c) $R_3 = \{(x,y) \in \mathbb{R} \times \mathbb{R} : y = 7x - 10\}$,
(d) $R_4 = \{(x,y) \in \mathbb{R} \times \mathbb{R} : y = x^2 + 2\}$,
(e) $R_5 = \{(x,y) \in \mathbb{R} \times \mathbb{R} : y = -4x^2 - 5\}$,
(f) $R_6 = \{(x,y) \in \mathbb{R} \times \mathbb{R} : y < x + 1\}$,
(g) $R_7 = \{(x,y) \in \mathbb{R} \times \mathbb{R} : y > 3x - 4\}$,
(h) $R_8 = \{(x,y) \in \mathbb{R} \times \mathbb{R} : y = \frac{2x}{x-2}\}$,
(i) $R_9 = \{(x,y) \in P \times P : y \text{ is the father of } x\}$
where P is the set of all people,
(j) $R_{10} = \{(x,y) \in P \times P : y \text{ is sibling of } x\}$,
(k) $R_{11} = \{(x,y) \in P \times P : y \text{ loves } x\}$.

14. Let

$R = \{(1,5),(2,2),(3,4),(5,2)\}$,
$S = \{(2,4),(3,4),(3,1),(5,5)\}$,
$T = \{(1,4),(3,5),(4,1)\}$.
Find

(a) $(R \circ S) \circ T$,   (b) $R \circ (S \circ T)$,
(c) $T \circ T$,   (d) $S \circ R$,
(e) $R \circ R$,   (f) $T \circ S$,
(g) $R \circ T$,   (h) $R \circ S$.

15. Find the composites for the relations defined in Exercise 11.

(a) $R_{10} \circ R_9$,   (b) $R_{11} \circ R_9$,
(c) $R_9 \circ R_9$,   (d) $R_8 \circ R_3$,
(e) $R_3 \circ R_8$,   (f) $R_8 \circ R_8$,
(g) $R_6 \circ R_6$,   (h) $R_2 \circ R_3$,
(i) $R_4 \circ R_5$,   (j) $R_1 \circ R_1$,
(k) $R_1 \circ R_2$,   (l) $R_2 \circ R_2$,
(m) $R_2 \circ R_4$,   (n) $R_6 \circ R_4$.

16. Let $A = \{1,2,3,4\}$. Give an example of relations R and S on A, such that

$$S \circ R \neq R \circ S \text{ and } (S \circ R)^{-1} \neq S^{-1} \circ R^{-1}.$$

17. Complete the proof of Theorem 3.2.27.

18. Let S and R be relations on a set A.

(a) Show by an example that one of the inclusions:

$$\text{Dom}(R) \subseteq \text{Dom}(S \circ R),$$

$$\text{Dom}(S \circ R) \subseteq \text{Dom}(R)$$

may be false.

(b) Which of these two statements must be true:

$\text{Rng}(S) \subseteq \text{Rng}(S \circ R)$ or $\text{Rng}(S \circ R) \subseteq \text{Rng}(S)$?

(c) Give an example of nonempty relations R and S on a set $A = \{1,2,3,4\}$, such that $R \circ S$ and $S \circ R$ are empty.

19. Prove that $(A \times B) \times C = A \times (B \times C)$ is false unless one of the sets is empty.

20. (a) Let R be a relation from A to B. For $a \in A$, define the vertical section of R at a by

$R_a = \{y \in B : (a,y) \in R\}$.
Prove that

$$\bigcup_{a \in A} R_a = \text{Rng}(R).$$

(b) Let R be a relation from A to B. For $b \in B$, define the horizontal section of R at b by the relation

$_bR = \{x \in A : (x,b) \in R\}$.
Prove that

$$\bigcup_{b \in B} {}_bR = \text{Dom}(R).$$

3.3   TYPE OF RELATIONS

Definition 3.3.1 Let R be a relation on a set A. R is called reflexive relation if

$$(x,x) \in R, \forall\, x \in A.$$

Example 3.3.2 Let $A = \{a,b,c,d\}$ and let R be a relation on A, defined by

$$R = \{(a,a),(a,c),(b,b),(b,d),(c,c),(c,d),(d,d)\}.$$

Then R is reflexive relation, since $(a,a),(b,b),(c,c),(d,d) \in R$.
If $T = \{(a,a),(a,c),(b,d),(c,c),(d,d)\}$ is a relation on A,
then T is not reflexive, since $(b,b) \notin T$.

Example 3.3.3 Let R_1, R_2, R_3, and R_4 be relations on $\mathbb{N}$ defined as follows:

$$R_1 = \{(x,y) \in \mathbb{N} \times \mathbb{N} : x < y\},$$
$$R_2 = \{(x,y) \in \mathbb{N} \times \mathbb{N} : x \leq y\},$$
$$R_3 = \{(x,y) \in \mathbb{N} \times \mathbb{N} : y \neq 0, y \,|\, x\},$$
$$R_4 = \{(x,y) \in \mathbb{N} \times \mathbb{N} : x + y = 5\}.$$

The relations R_2 and R_3 are reflexive but R_1 and R_4 are not reflexive. Why?

Example 3.3.4 Let X be arbitrary set and let R_1, R_2 be two relations on $P(X)$ defined as follows:

$$R_1 = \{(A,B) \in P(X) \times P(X) : A \subseteq B\},$$
$$R_2 = \{(A,B) \in P(X) \times P(X) : A \cap B = \phi\}.$$

R_1 is reflexive, since $A \subseteq A$, $\forall A \in P(X)$. But R_2 is not reflexive, since $A \cap A = A \neq \phi$.

Remark 3.3.5 If R is a reflexive relation on a set A, then the identity relation I_A is a subset of R. i.e.,

$$I_A \subseteq R.$$

Definition 3.3.6 Let R be a relation on a set A. R is called symmetric relation if

$$\forall x, y \in A, (x,y) \in R \Rightarrow (y,x) \in R.$$

Example 3.3.7 Let $A = \{1,2,3,4\}$ and let R be a relation on A defined by

$$R = \{(1,3),(4,2),(2,4),(3,1)\}.$$

Then R is symmetric.

Example 3.3.8 Let $A = \mathbb{N} - \{0\}$ and let R_1, R_2, and R_3 be relations on A defined by

$$R_1 = \{(x,y) \in A \times A : x \mid y\},$$
$$R_2 = \{(x,y) \in A \times A : x+y = 10\},$$
$$R_3 = \{(x,y) \in A \times A : x+3y = 10\}.$$

Then R_2 is symmetric but R_1 and R_3 are not symmetric. Why?

Example 3.3.9 Let X be a nonempty set and let R_1 and R_2 be relations on $P(X)$ defined by

$$R_1 = \{(A,B) \in P(X) \times P(X) : A \subset B\},$$
$$R_2 = \{(A,B) \in P(X) \times P(X) : A = X - B\}.$$

R_1 is not but R_2 is symmetric.

Theorem 3.3.10 Let R be a relation on a set A. Then R is symmetric iff

$$R = R^{-1}.$$

Proof Suppose R is symmetric, then

$$(x,y) \in R \Rightarrow (y,x) \in R \Leftrightarrow (x,y) \in R^{-1}.$$

Thus $R = R^{-1}$.
Conversely, suppose $R = R^{-1}$, then

$$(x,y) \in R \Rightarrow (x,y) \in R^{-1}$$
$$\Rightarrow (y,x) \in R.$$

Hence, R is symmetric.

Definition 3.3.11 Let R be a relation on a set A. R is called transitive relation if

$$\forall\, x,y,z \in A, [(x,y) \in R \wedge (y,z) \in R] \Rightarrow (x,z) \in R.$$

Example 3.3.12 Let R be a relation on $\mathbb{N}$ defined by

$$R = \{(x,y) in \mathbb{N} \times \mathbb{N} : x < y\}.$$

R is transitive, since

$$x < y \wedge y < z \Rightarrow x < z.$$

That is

$$(x,y) \in R \wedge (y,z) \in R \Rightarrow (x,z) \in R.$$

Example 3.3.13 Let $A = \{1,2,3\}$. Let R_1, R_2, and R_3 be relations on A defined by

$$R_1 = \{(1,2),(2,2)\},$$
$$R_2 = \{(1,1)\},$$
$$R_3 = \{(1,2),(2,3),(1,3),(2,1),(1,1)\}.$$

R_1 and R_2 are transitive. Why? But R_3 is not transitive. Why?

Definition 3.3.14 Let R be a relation on a set A. R is called antisymmetric relation if

$$(x,y) \in R \wedge (y,x) \in R \Rightarrow x = y.$$

Remark 3.3.15 The statement: "R is not symmetric" does not mean that R is antisymmetric.

Example 3.3.16 Let X be an arbitrary set and let R be a relation on $P(X)$ defined by

$$R = \{(A,B) \in P(X) \times P(X) : A \subset B\}.$$

R is antisymmetric, since

$$A \subseteq B \wedge B \subseteq A \Rightarrow A = B.$$

Example 3.3.17 Let T be a relation on $\mathbb{N}$ defined by

$$T = \{(x,y) \in \mathbb{N} \times \mathbb{N} : x \leq y\}.$$

T is antisymmetric, since

$$x \leq y \wedge y \leq x \Rightarrow x = y.$$

Example 3.3.18 Let $A = \{1,2,3,4\}$, and let R be a relation on A defined by

$$R = \{(1,3),(3,1),(3,3),(3,2)\}.$$

R is not antisymmetric, since $(1, 3), (3, 1) \in R$, but $1 \neq 3$.

Theorem 3.3.19 Let R be a relation on a set A. Then R is antisymmetric iff

$$R \cap R^{-1} = I_A.$$

Proof Suppose that R is antisymmetric and let $(x,y) \in R \cap R^{-1}$. Then

$$
\begin{aligned}
(x,y) \in R \cap R^{-1} &\Leftrightarrow (x,y) \in R \wedge (x,y) \in R^{-1} \\
&\Leftrightarrow (x,y) \in R \wedge (y,x) \in R \\
&\Leftrightarrow x = y \text{(since } R \text{ is antisymmetric)} \\
&\Leftrightarrow (x,y) \in I_A.
\end{aligned}
$$

Thus $R \cap R^{-1} = I_A$.
Conversely, suppose that $R \cap R^{-1} = I_A$, and let

$$(x,y) \in R \wedge (y,x) \in R.$$

Then $(x,y) \in R \wedge (x,y) \in R^{-1}$. Hence $(x,y) \in R \cap R^{-1}$. Therefore $(x,y) \in I_A$ and thus $x = y$. This proves that R is antisymmetric.

Exercises

1. Let S be a relation on a set X. Prove that

   (a) S is transitive iff $S \circ S = S$.
   (b) if S is reflexive and transitive, then

   $$S \circ S = S.$$

2. Let S and T be transitive relations on a set X. Is $S \cap T$ transitive?

3. Consider the following relations on the set $A = \{0,1,2,\ldots,11\}$:

   (a) $R_1 = \{(x,y) : y - x \in A\}$,
   (b) $R_2 = \{(x,y) : y - x \in A \text{ and } y - x \neq 0\}$,
   (c) $R_3 = \{(x,y) : 1 < x - y\}$,
   (d) $R_4 = \{(x,y) : y - x = 1\}$.
   Determine the type of relation in each case.

4. Let R be a relation on a set X. Prove that $R \cup R^{-1}$ is the smallest symmetric relation, such that $R \subset R \cup R^{-1}$, and $R \cap R^{-1}$ is the greatest symmetric relation, such that $R \cap R^{-1} \subset R$.

5. Let G and H be relations on a set A. Suppose that G is reflexive and H is both reflexive and transitive. Prove that

   $$G \subseteq H \Leftrightarrow G \circ H = H.$$

6. Let H be a reflexive relation and G be an arbitrary relation on a set A. Prove that

   $$G \subseteq H \circ G \text{ and } G \subseteq G \circ H.$$

7. Give an example, if possible, of a set A, such that every relation on A is symmetric.

8. Give an example, if possible, of a relation R on a set A, such that R is symmetric and antisymmetric in the same time.

9. Let R and R' be relations on a set A, such that R is reflexive. Prove that $R \cup R'$ is reflexive.

10. Let R and R' be relations on a set A. Show that the following statements are false:
(a) If R is antisymmetric and R' is also antisymmetric, then $R \cup R'$ is antisymmetric. (b) If R is transitive and R' is also transitive, then $R \cup R'$ is transitive.

3.4   EQUIVALENCE RELATIONS AND EQUIVALENCE CLASSES

Definition 3.4.1 Let R be a relation on a set A. Then R is called an equivalence relation iff R is reflexive, symmetric, and transitive.

Example 3.4.2 Let R_1 and R_2 be relations on $\mathbb{R}$ defined by

$$R_1 = \{(x,y) \in \mathbb{R} \times \mathbb{R} : x = y\},$$
$$R_2 = \{(x,y) \in \mathbb{R} \times \mathbb{R} : x < y\}.$$

R_1 is an equivalence relation on $\mathbb{R}$, since
1. $x = x$, $\forall x \in \mathbb{R}$, that is $(x,x) \in R_1$, $\forall x \in R$,
2. $x = y \Rightarrow y = x$, that is $(x,y) \in R_1 \Rightarrow (y,x) \in R_1$,
3. $[x = y \wedge y = z] \Rightarrow x = z$, that is
$[(x,y) \in R_1 \wedge (y,z) \in R_1] \Rightarrow (x,z) \in R_1$.
Therefore R_1 is reflexive, symmetric, and transitive on $\mathbb{R}$.
Thus R_1 is an equivalence relation on $\mathbb{R}$.
But R_2 is not an equivalence relation, since it is neither reflexive nor symmetric.

Example 3.4.3 Let X be a nonempty set and let S and T be relations on $P(X)$ defined by

$$S = \{(A,B) \in P(X) \times P(X) : A = B\}$$
$$T = \{(A,B) \in P(X) \times P(X) : A \subseteq B\}$$

S is an equivalence relation. Why? But T is not an equivalence relation. Why?

Example 3.4.4 Let A be the set of all the lines on a plane and let R and S be relations on A defined by

$$R = \{(x,y) \in A \times A : x // y\},$$
$$S = \{(x,y) \in A \times A : x \lfloor y\}.$$

R is an equivalence relation, but S is not. Why?

Example 3.4.5 Let $A = \mathbb{R} \times \mathbb{R}$ and let R be a relation on A, defined by

$$R = \{(a,b),(c,d) \in A \times A : a+b = c+d\}.$$

"Below we show that R is an equivalence relation on A."

1. If $(a,b) \in A$, then $a+b = a+b$. Therefore $((a,b),(a,b)) \in R$, $\forall$ $(a,b) \in A$. Thus R is a reflexive relation.

2. If $((a,b),(c,d)) \in R$, then $a+b = c+d$. But

$$a+b = c+d \Rightarrow c+d = a+b.$$

Therefore $((c,d),(a,b)) \in R$. Thus R is a symmetric relation.

3. If $((a,b),(c,d)) \in R \wedge ((c,d),(e,f)) \in R$, then

$$a+b = c+d \wedge c+d = e+f; i.e., a+b = e+f.$$

Hence $((a,b),(e,f)) \in R$. Thus R is transitive relation.

Theorem 3.4.6 If R and S are equivalence relations on a set A. Then $R \cap S$ is an equivalence relation.

Proof

1. Since R and S are reflexive, then

$$\forall x \in A, (x,x) \in R \wedge (x,x) \in S.$$

Thus $(x,x) \in R \cap S$, $\forall x \in A$. Hence $R \cap S$ is a reflexive relation on A.

2. Suppose $(x,y) \in R \cap S$, then

$$
\begin{aligned}
(x,y) \in R \cap S &\Rightarrow (x,y) \in R \wedge (x,y) \in S \\
&\Rightarrow (y,x) \in R \wedge (y,x) \in S \text{ (since } R \text{ and } S \text{ are symmetric)} \\
&\Rightarrow (y,x) \in R \cap S.
\end{aligned}
$$

Thus $R \cap S$ is symmetric.

3. Suppose $(x,y) \in R \cap S \wedge (y,z) \in R \cap S$, then

$$[(x,y) \in R \wedge (x,y) \in S] \wedge [(y,z) \in R \wedge (y,z) \in S].$$

That is $[(x,y) \in R \wedge (y,z) \in R] \wedge [(x,y) \in S \wedge (y,z) \in S]$.
This implies $(x,z) \in R \wedge (x,z) \in S$, since R and S are transitive.
Hence, $(x,z) \in R \cap S$. Thus, $R \cap S$ is transitive relation. Therefore, $R \cap S$ is an equivalence relation.

Theorem 3.4.7 If R is an equivalence relation on a set A. Then

$$R \circ R = R.$$

Proof Suppose that $(x, z) \in R \circ R$, then $\exists\, y \in A$, such that

$$(x, y) \in R \wedge (y, z) \in R.$$

Since R is a transitive relation, then

$$(x, y) \in R \wedge (y, z) \in R \Rightarrow (x, z) \in R.$$

Thus

$$R \circ R \subseteq R. \qquad (3.11)$$

Now, suppose that $(x, y) \in R$. Then $(x, x) \in R$, since R is reflexive. Thus, $\exists\, x \in A$, such that

$$(x, x) \in R \wedge (x, y) \in R.$$

This implies $(x, y) \in R \circ R$ and hence

$$R \subseteq R \circ R. \qquad (3.12)$$

From Equations (3.11) and (3.12), we conclude

$$R \circ R = R.$$

Definition 3.4.8 Let R be an equivalence relation on a nonempty set A and let $a \in A$. The equivalence class of a, determined by R, is the set

$$[a] = a/R = \{b \in A : (a, b) \in R\}.$$

Example 3.4.9 Let $A = \{1, 2, 3, 4\}$ and let R be a relation on A defined by

$$R = \{(1,1), (2,2), (3,3), (4,4), (1,3), (3,1)\}.$$

It is clear that R is an equivalence relation on A and the equivalence classes are

$$[1] = \{x \in A : (x, 1) \in R\} = \{1, 3\},$$
$$[2] = \{x \in A : (x, 2) \in R\} = \{2\},$$
$$[3] = \{x \in A : (x, 3) \in R\} = \{1, 3\} = [1],$$
$$[4] = \{x \in A : (x, 4) \in R\} = \{4\}.$$

Since $[1] = [3]$, then the equivalence classes are

$$[1], [2], [4].$$

Theorem 3.4.10 (Properties of equivalence classes) Let R be an equivalence relation on a set A and let $(a,b) \in A$. Then

(a) $a \in [a]$,
(b) if $b \in [a]$, then $[a] = [b]$,
(c) $[a] = [b]$ iff $(a,b) \in R$,
(d) if $[a] \cap [b] \neq \phi$, then $[a] = [b]$.

Proof (a) By definition of $[a]$, we have

$$[a] = \{x \in A : (x,a) \in R\}.$$

Since R is a reflexive relation, then

$$(a,a) \in R, \forall\, a \in A.$$

Thus $a \in [a]$.

(b) Suppose that $b \in [a]$ and let $x \in [b]$. Then $x \in [b] \Rightarrow (x,b) \in R$.

Since $b \in [a]$, then $(b,a) \in R$. Hence
$(x,b) \in R \wedge (b,a) \in R \Rightarrow (x,a) \in R$ (since R is transitive),
$$\Rightarrow x \in [a].$$

Thus

$$[b] \subseteq [a]. \tag{3.13}$$

Now, suppose that $y \in [a]$. Then
$$y \in [a] \Rightarrow (y,a) \in R.$$
Hence

$$b \in [a] \Rightarrow (b,a) \in R$$
$$\Rightarrow (a,b) \in R,$$

since R is symmetric. Consequently

$$(y,a) \in R \wedge (a,b) \in R \Rightarrow (y,b) \in R$$
$$\Rightarrow y \in [b],$$

since R is transitive. Therefore

$$[a] \subseteq [b]. \tag{3.14}$$

Thus, it follows from Equations (3.13) and (3.14) that

$$[a] = [b].$$

(c) Firstly, suppose that $[a] = [b]$. Then by part (a), we have

$$a \in [a].$$

Since $[a] = [b]$, then

$$a \in [a] \Rightarrow a \in [b]$$
$$\Rightarrow (a,b) \in R.$$

Conversely, suppose that $x \in [a]$. Then,

$$x \in [a] \Rightarrow (x,a) \in R.$$

This implies $(x,a) \in R \wedge (a,b) \in R$. Hence,

$$(x,a) \in R \wedge (a,b) \in R \Rightarrow (x,b) \in R$$
$$\Rightarrow x \in [b].$$

Thus,

$$[a] \subseteq [b]. \tag{3.15}$$

By the same manner, we prove that

$$[b] \subseteq [a]. \tag{3.16}$$

From Equations (3.15) and (3.16), we conclude

$$[a] = [b].$$

(d) Suppose that $[a] \cap [b] \neq \phi$, and let $x \in [a] \cap [b]$. Then

$$x \in [a] \cap [b] \Rightarrow x \in [a] \wedge x \in [b]$$
$$\Rightarrow (x,a) \in R \wedge (x,b) \in R$$
$$\Rightarrow (x,b) \in R \wedge (x,a) \in R$$
$$\Rightarrow (b,x) \in R \wedge (x,a) \in R(\text{since } R \text{ is symmetric})$$
$$\Rightarrow (b,a) \in R(\text{since } R \text{ is transitive})$$
$$\Rightarrow (a,b) \in R(\text{since } R \text{ is symmetric})$$
$$\Rightarrow [a] = [b](\text{see part (c)})$$

Definition 3.4.11 Let $\{A_i\}_{i \in I}$ be a family of nonempty subsets of A. Then $\{A_i\}_{i \in I}$ is called partition of A iff

1. $\forall i, j \in I, A_i \cap A_j = \phi \wedge A_i = A_j$.
2. $A = \bigcup_{i \in I} A_i$.

Example 3.4.12 Let $A = \mathbb{Z}$ and let $X = \mathbb{Z}_e$ and $Y = \mathbb{Z}_o$. It is clear that X and Y are nonempty subsets of A and $X \cap Y = \phi$, also $A = X \cup Y$. Therefore, the set $\{X,Y\}$ is a partition of A.

Theorem 3.4.13 Let R be an equivalence relation on a set A and let $\{[a]\}_{a \in A}$ be the family of all the equivalence classes of R in A. Then $\{[a]\}_{a \in A}$ is a partition of A.

Proof It is clear that $[a] \subseteq A$, $\forall a \in A$. Since R is a reflexive relation, then $(a,a) \in R$ and $a \in [a]$ (by Theorem 3.4.10 part (a)). Thus $[a] \neq \phi$, $\forall a \in A$.

Now, suppose that $\exists a,b \in A$, such that $[a] \cap [b] \neq \phi$. Then $[a] = [b]$ (by Theorem 3.4.10 part (d)).

Finally, we must show that $A = \bigcup_{a \in A} [a]$.

Suppose that $x \in A$, then $x \in [x]$ (by Theorem 3.4.10 part (a)). Thus, $x \in \bigcup_{a \in A} [a]$, that is $A \subseteq \bigcup_{a \in A} [a]$ and it is clear that $\bigcup_{a \in A} [a] \subseteq A$.

Thus, $A = \bigcup_{a \in A} [a]$. Hence, $\{[a]\}_{a \in A}$ is a partition of A.

Theorem 3.4.14 If A is a nonempty set and $\{A_i\}_{i \in I}$ is a partition of A, then there exists an equivalence relation R on A, such that $\{A_i\}_{i \in I}$ are the equivalence classes of R.

Proof Let R be a relation on A defined by

$$R = \{(x,y) \in A \times A : \exists A_i, \text{ such that } x,y \in A_i\}.$$

Below we show that R is an equivalence relation.

1. Let $x \in A$. Since $\{A_i\}_{i \in I}$ is a partition of A, then

$$A = \bigcup_{i \in I} A_i.$$

So, $\exists A_i$, such that $x \in A_i$. Then

$$x \in A_i \wedge x \in A_i \Rightarrow (x,x) \in R.$$

Thus R is a reflexive relation.

2. Suppose that $(x,y) \in R$. Then $\exists A_i$, such that $x,y \in A_i$. That is $\exists A_i$, $y \in A_i \wedge x \in A_i$. Hence $(y,x) \in R$. Hence R is symmetric.

3. Suppose that $(x,y) \in R \wedge (y,z) \in R$. Then $\exists A_i, A_j$, such that $x,y \in A_i \wedge y, z \in A_j$. Thus $y \in A_i \cap A_j$. Hence $A_i \cap A_j \neq \phi$, and $\{A_i\}_{i \in I}$ is a partition of the set A. Consequently $A_i = A_j$.
So, $\exists A_i$, such that $x \in A_i \wedge z \in A_i$ and from the definition of R, we have $(x,z) \in R$. Therefore, R is a transitive relation.
Thus R is an equivalence relation.

Now, we show that every subset A_i in $\{A_i\}_{i \in I}$ is an equivalence class for R. As $A_i \neq \phi$, $\forall i \in I$, then A_i has at least one element x, so $[x]$ is an equivalence class for R.

We claim that $Ai = [x]$. Suppose that $y \in [x]$. Then $(y,x) \in R$, and since $x \in A_i$, then by definition of R, $y \in A_i$. That is

$$y \in [x] \Rightarrow y \in A_i.$$

Then

$$z \in A_i \wedge x \in A_i$$
$$\Rightarrow (z,x) \in R$$
$$\Rightarrow z \in [x].$$

Consequently,

$$z \in A_i \Rightarrow z \in [x],$$

that is $A_i \subseteq [x]$. Hence $A_i = [x]$.

Example 3.4.15 Let $A = \{1,3,5,7,9\}$, $X = \{1,3\}$, $Y = \{5,7\}$, and $Z = \{9\}$. Then by Theorem 4.2.47, there exists an equivalence relation R, such that the equivalence classes of R form the partition $\{X,Y,Z\}$.

The relation R is

$$R = I_A \cup \{(1,3),(3,1),(5,7),(7,5)\}.$$

It is easy to prove the following properties:

1. The set $\{X,Y,Z\}$ is a partition of the set A.
2. R is an equivalence relation on A.

The equivalence classes of R are

$$[1] = [3], [5] = [7], [9].$$

We note that

$$X = [1] = [3],$$
$$Y = [5] = [7],$$
$$Z = [9].$$

Example 3.4.16 Let $A = \mathbb{Z}$, $X = \mathbb{Z}_e$, and $Y = \mathbb{Z}_o$. The set $\{X,Y\}$ is a partition of A. Consider

$$R = \{(x,y) \in A \times A : (x-y) \text{ is an even integer}\}.$$

The relation R is an equivalence relation on A and the equivalence classes of R in A are

$$[0] \text{ and } [1], \text{ where } X = [0], Y = [1].$$

3.5 CONGRUENCE

Definition 3.5.1 Let n be a positive integer. If a and b are integers, we say that a is congruent to b modulo n, and we write

$$a \equiv b \pmod{n} \Leftrightarrow n \text{ divides } (a-b).$$

Thus "$\equiv$" defines a relation on $\mathbb{Z}$.

For example,

$$21 \equiv 1 \pmod 5$$

because $5 \mid (21 - 1)$. Likewise,

$$35 \equiv -7 \ (\text{mod } 6)$$
$$19 \equiv 82 \ (\text{mod } 9).$$

On the other hand, $67 \equiv 5 \ (\text{mod } 11)$ is false, since

$$11 \mid (67 - 5).$$

If $a \equiv b \ (\text{mod } n)$ is false, then we write

$$a \equiv b \ (\text{mod } n).$$

Thus $67 \equiv 5 \ (\text{mod } 11)$ and $-17 \equiv 2 \ (\text{mod } 4)$.

Theorem 3.5.2 For a fixed positive integer n, the relation of congruence modulo n is an equivalence relation on $\mathbb{Z}$.

Proof

1. As

$$n \mid 0 \Rightarrow n \mid (a - a)$$
$$\Rightarrow a \equiv a \ (\text{mod } n),$$

then $\equiv$ is a reflexive relation on $\mathbb{Z}$.

2. As

$$a \equiv b \ (\text{mod } n) \Leftrightarrow n \mid (a - b)$$
$$\Leftrightarrow n \mid -(b - a)$$
$$\Leftrightarrow n \mid (b - a).$$

then $b \equiv a \ (\text{mod } n)$. Thus $\equiv$ is a symmetric relation on $\mathbb{Z}$.

3. We show that $\equiv$ is transitive. Suppose

$$a \equiv b \ (\text{mod } n) \wedge b \equiv c \ (\text{mod } n).$$

Then $n \mid (a - b) \wedge n \mid (b - c)$.
Therefore, there exists integers r, s, such that

$$a - b = rn \wedge b - c = sn.$$

Then $a - c = (r + s)n$, where $r + s$ is an integer. Thus $n \mid (a - c)$, so $a \equiv c \ (\text{mod } n)$. Therefore $\equiv$ is a transitive relation on $\mathbb{Z}$. As $\equiv$ is reflexive, symmetric, and transitive relation on $\mathbb{Z}$, then $\equiv$ is an equivalence relation on $\mathbb{Z}$.

By the preceding theorem, congruence (modulo n) divides $\mathbb{Z}$ into equivalence classes for any fixed positive integer n. For example, if $n = 3$, then

$$
\begin{aligned}
[7] &= \{x \in \mathbb{Z} : x \equiv 7 \ (\mathrm{mod} \ 3)\} \\
&= \{x \in \mathbb{Z} : 3 \mid (x - 7)\} \\
&= \{x \in \mathbb{Z} : x - 7 = 3k, \ k \in \mathbb{Z}\} \\
&= \{x \in \mathbb{Z} : x = 7 + 3k, \ k \in \mathbb{Z}\} \\
&= \{\ldots, -2, 1, 4, 7, 10, 13, 16, 19, \ldots\}.
\end{aligned}
$$

By the same computation, we find

$$
[1] = \{\ldots, -2, 1, 4, 7, 10, 13, 16, 19, \ldots\}.
$$

Thus $[1] = [7]$. Likewise

$$
[0] = [3] = [6] = [9] = \ldots
$$

We see that the equivalence relation congruence (modulo 3) divides $\mathbb{Z}$ into exactly 3 distinct equivalence classes, namely $[0], [1]$, and $[2]$. A similar result is true in general.

Theorem 3.5.3 If n is a positive integer, then congruence (modulo n) divides $\mathbb{Z}$ into exactly n distinct equivalence classes, namely

$$
[0], [1], [2], \ldots, [n - 1].
$$

In fact, if r is an integer with $0 \leq r < n$, then $x \in [r]$ iff r is the remainder when x is divided by n using the division algorithm.

Proof Let $x = nq + r, \ 0 \leq r < n$. Then

$$
x - r = nq, \text{ and so } n \mid (x - r).
$$

Thus

$$
x \equiv r \ (\mathrm{mod} \ n), \text{ and so } x \in [r].
$$

Conversely, let $x \in [r]$, where $0 \leq r < n$. Then

$$
x \equiv r \ (\mathrm{mod} \ n) \text{ and so } n \mid (x - r).
$$

Let $x - r = nk, \ k \in \mathbb{Z}$. Then

$$
x = nk + r, \ 0 \leq r < n.
$$

By the uniqueness of the quotient and the remainder in the division algorithm, r must be the remainder, when x is divided by n.

Theorem 3.5.4 Suppose that a, b, c, d, and n, are integers with $n > 1$. If

$$
a \equiv b \ (\mathrm{mod} \ n) \text{ and } c \equiv d \ (\mathrm{mod} \ n),
$$

then

(a) $a + c \equiv b + d \ (\mathrm{mod} \ n)$,
(b) $ac \equiv bd \ (\mathrm{mod} \ n)$.

Proof

(a) As

$$a \equiv b \pmod{n} \Leftrightarrow a = b + sn, \ s \in \mathbb{Z}$$

and

$$c \equiv d \pmod{n} \Leftrightarrow c = d + tn, \ t \in \mathbb{Z}$$
$$\Leftrightarrow a + c = b + d + (s+t)n$$
$$\Leftrightarrow a + c = b + d + kn, \ k \in \mathbb{Z},$$

then $a + c \equiv b + d \pmod{n}$.

(b) By hypothesis, we have

$$a \equiv b \pmod{n} \Rightarrow n \mid (a - b)$$

and

$$c \equiv d \pmod{n} \Rightarrow n \mid (c - d).$$

Then

$$n \mid (a - b)c + (c - d)b; i.e., n \mid ac - bd.$$

This means that $ac \equiv bd \pmod{n}$.

Example 3.5.5 Suppose that $a \equiv 2 \pmod 7$ and $b \equiv 5 \pmod 7$. We will determine the remainder, when $45a + b$ is divided by 7.
For this purpose, we write

$$45 = 7.6 + 3; \ i.e., 45 \equiv 3 \pmod 7,$$

and

$$a \equiv 2 \pmod 7.$$

Multiplying these congruence's, we obtain

$$45a \equiv 6 \pmod 7.$$

Adding $b \equiv 5 \pmod 7$ to the latter congruence, we get

$$45a + b \equiv 11 \pmod 7.$$

Hence

$$45a + b \in [11] = [4].$$

Exercises

1. When is a relation R on a set A not reflexive?

2. Let $A = \{1,2,3,4\}$ and $R = \{(1,10),(1,3),(2,2),(3,1),(4,4)\}$. Is R reflexive?

3. Let R,S,T, and H be relations on $\mathbb{N}$, defined by

   $R = \{(x,y) \in \mathbb{N} \times \mathbb{N} : x \mid y\}$,
   $S = \{(x,y) \in \mathbb{N} \times \mathbb{N} : x \leq y\}$,
   $T = \{(x,y) \in \mathbb{N} \times \mathbb{N} : x \text{ and } y \text{ are relatively prime}\}$,
   $H = \{(x,y) \in \mathbb{N} \times \mathbb{N} : x+y = 10\}$.
   State whether or not each of these relations is reflexive.

4. Let $B = \{1,2,3\}$. Consider the following relations on B:

   $G = \{(1,2),(3,2),(2,2),(2,3)\}$,
   $H = \{(1,20),(2,3),(1,3)\}$,
   $K = \{(1,1),(2,2),(2,3),(3,2),(3,3)\}$,
   $L = \{(1,2)\}$,
   $P = B \times B$.
   State whether or not each of these relations is reflexive.

5. When is a relation R on a set A not symmetric?

6. Is there a set A, in which every relation is symmetric?

7. State whether or not each relation in Exercise 3 is symmetric.

8. State whether or not each relation in Exercise 4 is symmetric.

9. State whether or not each relation in Exercise 3 is transitive.

10. State whether or not each relation in Exercise 4 is transitive.

11. State whether or not each relation in Exercise 3 is antisymmetric.

12. Let $C = \{i,-1,-i,1\}$, where $i^2 = -1$, and let R be a relation on C defined by

    $$R = \{(x,y) \in C \times C : xy = \pm 1\}.$$

    Show that R is an equivalence relation on C. Give the partition of C associated with R.

13. List the ordered pairs in the equivalence relation on

    $$A = \{1,2,3,4,5\}$$

    associated with the following partitions

    (a) $\{\{1,2\},\{3,4,5\}\}$,
    (b) $\{\{1\},\{2\},\{3\},\{4\},\{5\}\}$,
    (c) $\{\{2,3,4,5\},\{1\}\}$.

14. Let $C = \{i, -1, -i, 1\}$, where $i^2 = -1$, and let R be a relation on $C \times C$ defined by

$$R = \{((x,y),(u,v)) \in (C \times C) \times (C \times C) : xy = uv\}.$$

Show that R is an equivalence relation, and give the partition of $C \times C$ associated with R.

15. Describe the partition for each of the following relation:

   (a) $R = \{(x,y) \in \mathbb{R} \times \mathbb{R} : x - y \in \mathbb{Z}\}$,
   (b) $S = \{(x,y) \in \mathbb{Z} \times \mathbb{Z} : x + y \text{ is even}\}$,
   (c) $T = \{(x,y) \in \mathbb{R} \times \mathbb{R} : \sin x = \sin y\}$.

16. Describe the equivalence relation on each of the following sets with the given partition:

   (a) $\mathbb{N}.\{\{1\}, \{2,3\}, \{4,5,6,7\}, \{8,9,10,11,12,13,14,15\}, \ldots\}$,
   (b) $\mathbb{Z}.\{\ldots, \{-2\}, \{-1\}, \{0\}, \{1\}, \{2\}, \{3,4,5,\ldots\}\}$,
   (c) $\mathbb{R}.\{(-\infty,0), \{0\}, (0,\infty)\}$,
   (d) $\mathbb{Z}.\{A,B\}$, where $A = \{x \in \mathbb{Z} : x < 3\}$ and $B = \mathbb{Z} - A$.

17. Let A be the set of all people and let R be a relation on A defined by

$$R = \{(x,y) \in A \times A : x \text{ is the brother of } y\}.$$

Is R reflexive, symmetric or transitive?

18. Let A be the set of all people and let R be a relation on A defined by

$$R = \{(x,y) \in A \times A : x \text{ is the mother of } y\}.$$

Is R reflexive or symmetric or transitive?

19. Let $A = \mathbb{Z}$ and let R be a relation on A defined by

$$R = \{(x,y) \in A \times A : |x - y| < 2\}.$$

Is R reflexive, symmetric or transitive?

20. Let $A = \mathbb{Z}$ and let R be a relation on A defined by

$$R = \{(x,y) \in A \times A : x - y \text{ is even or } xy > 5\}.$$

Is R reflexive, symmetric or transitive?

21. Let $A = \mathbb{Z}$ and let R be a relation on A defined by

$$R = \{(x,y) \in A \times A : xy = 1\}.$$

Is R reflexive? Symmetric? Transitive?

22. Let $A = R$ and let R be a relation on A defined by

$$R = \{(x,y) \in A \times A : x - y \in \mathbb{Q}\}.$$

Is R reflexive, symmetric or transitive?

23. Let $A = R$ and let T be a relation on A defined by

$$T = \{(x,y) \in A \times A : xy = 0\}.$$

Is R reflexive, symmetric or transitive?

24. Determine the equivalence classes of the equivalence relation R on A:

(a) $A = \{1,2,3,4,5,6\}$, $R = \{(x,y) \in A \times A : x - y \in \{-3,0,3\}\}$,
(b) $A = \{1,2,3,4,5,6\}$, $R = \{(x,y) \in A \times A : |3 - x| = |3 - y|\}$,
(c) $A = \{n \in \mathbb{Z} : n \neq 0\}$, $R = \{(x,y) \in A \times A : xy > 0\}$,
(d) $A = \mathbb{R}$, $R = \{(x,y) \in A \times A : x^2 = y^2\}$.

25. Prove that if R is an equivalence relation on a set A, then

$$(x,y) \in R \text{ iff } [x] = [y].$$

26. Let $A = \{1,2,3\}$. Give an example of a relation on A with the following specified properties:

(a) Reflexive and symmetric, but not transitive,
(b) Symmetric and transitive, but not reflexive,
(c) Transitive, but not reflexive or symmetric.

27. Let R and S be relations on a set A. State whether each of the following statements is true or false:

(a) If R is symmetric, then R^{-1} is symmetric.
(b) If R is transitive and S is transitive, then $R \cup S$ is transitive.
(c) If R is reflexive and S is reflexive, then $R \cap S$ is reflexive.

28. What type of relation is R if

(a) $R \cap R^{-1} = \phi$?
(b) $R = R^{-1}$?

29. Write all the equivalence relations on the set

$$A = \{0,1,2\}.$$

30. Is ϕ an equivalence relation?

31. Let R and S be equivalence relations on a set A. Prove that $S \circ R$ is an equivalence relation on A iff $S \circ R = R \circ S$.

32. Let R be a relation on $\mathbb{R} \times \mathbb{R}$ defined by

$$R = \{(a,b),(c,d)) : b-a=d-c\}.$$

Prove that R is an equivalence relation on $\mathbb{R} \times \mathbb{R}$.

33. Let S be an equivalence relation on a set X and T be an equivalence relation on a set Y. Define the relation R on $X \times Y$ by

$$(x,y)R(z,w) \Leftrightarrow (x,z) \in S \wedge (y,w) \in T.$$

Prove that R is an equivalence relation on $X \times Y$.

34. Let R be a reflexive and transitive relation on a set A. Let $\sim$ be a relation on A defined by

$$a \sim b \text{ iff } aRb \wedge bRa.$$

Prove that $\sim$ is an equivalence relation on A.

35. Let R and S be relations on a set A. State whether each of the following statements is true or false:

(a) If R and S are antisymmetric, then $R \cup S$ is antisymmetric.
(b) If R and S are antisymmetric, then
$R \cap S$ is antisymmetric.
(c) If R is antisymmetric, then R^{-1} is antisymmetric.

36. Calculate the equivalence classes for the relation of

(a) congruence modulo 1,
(b) congruence modulo 5,
(c) congruence modulo 7,
(d) congruence modulo 8.

37. Prove that if

(a) $x \equiv y$ (modulo n), then $[x] = [y]$,
(b) $[x] = [y]$, then $x \equiv y$ (mod n),
(c) $[x] \cap [y] \neq \phi$, then $[x] = [y]$.

38. Let R and S be relations on $\mathbb{N}$ defined by

$R = \{(x,y) \in \mathbb{N} \times \mathbb{N} : 2 \mid (x+y)\}$,
$S = \{(x,y) \in \mathbb{N} \times \mathbb{N} : 3 \mid (x+y)\}$.
Prove that R is an equivalence relation but S is not.

3.6   PARTIAL AND TOTAL ORDERED RELATIONS

Definition 3.6.1 Let R be a relation on a set A. Then R is called a partial order (or partial ordering) relation for A if R is reflexive, antisymmetric, and transitive. The set A is called a partially ordered set, or poset.

Example 3.6.2 Let R be a relation on $\mathbb{Z}$ defined by

$$R = \{(x,y) \in \mathbb{Z} \times \mathbb{Z} : x \leq y\},$$

R is a partial order relation on $\mathbb{Z}$. Why ?

Example 3.6.3 Let S be a relation on $\mathbb{Z}$ defined by

$$S = \{(x,y) \in \mathbb{Z} \times \mathbb{Z} : 3 \mid (x-y),$$

S is not a partial order relation on $\mathbb{Z}$, because S is not antisymmetric. For example,

$$(8,2) \in S \wedge (2,8) \in S, \text{ but } 2 \neq 8.$$

Example 3.6.4 Let X be a nonempty set and R be a relation on $P(X)$ defined by

$$R = \{(A,B) \in P(X) \times P(X) : A \subseteq B\},$$

R is a partial order relation on $P(X)$.

Remark 3.6.5 Let $<$ be the usual "strict" inequality on the natural numbers (or, integers, or rational numbers, or real numbers). Then it is not a partial ordering because it is not reflexive.

Example 3.6.6 Let S be a relation on $\mathbb{R}$ defined by

$$S = \{(x,y) \in \mathbb{R} \times \mathbb{R} : x < y\}.$$

Then this relation S is not a partial ordering on $\mathbb{R}$.

Theorem 3.6.7 If R is a partial order relation on a set A, then R^{-1} is a partial order relation on A.

Proof

1. Since R is reflexive; i.e., $(x,x) \in R, \forall x \in A$, then

$$\forall x \in A, (x,x) \in R \Rightarrow (x,x) \in R^{-1}.$$

Thus R^{-1} is reflexive.

2. Since R is antisymmetric; i.e.,

$$\forall x,y \in A, (x,y) \in R \wedge (y,x) \in R \Rightarrow x = y,$$

then
$$\forall x,y \in A, (y,x) \in R^{-1} \wedge (x,y) \in R^{-1} \Rightarrow x = y.$$

Thus R^{-1} is antisymmetric.

3. Since R is transitive; i.e.,

$$\forall x, y, z \in A, (x,y) \in R \wedge (y,z) \in R \Rightarrow (x,z) \in R,$$

then

$$\forall x, y, z \in A, (y,x) \in R^{-1} \wedge (z,y) \in R^{-1} \Rightarrow (z,x) \in R^{-1}$$

Thus R^{-1} is transitive.

As R^{-1} is reflexive, antisymmetric, and transitive, then R^{-1} is a partial order relation on A.

Theorem 3.6.8 Let R be a relation on a set A. Then R is a partial order relation on A iff $R \cap R^{-1} = I_A \wedge R \circ R = R$.

Proof Suppose that $R \cap R^{-1} = I_A \wedge R \circ R = R$. We prove that R is a partial order relation. First we see that

$$\forall a \in A, (a,a) \in I_A,$$

by definition of I_A. Then

$$\forall a \in A, (a,a) \in R \cap R^{-1}.$$

That is, $\forall a \in A, (a,a) \in R$.

Thus R is reflexive relation.

Now, suppose $(a,b) \in R \wedge (b,a) \in R$, but $(a,b) \in R \Rightarrow (b,a) \in R^{-1}$.

Then

$$(b,a) \in R \cap R^{-1} = I_A; i.e., (b,a) \in I_A.$$

Hence $a = b$. This implies that R is symmetric.

Finally, suppose $(a,b) \in R \wedge (b,c) \in R$. Since $R \circ R = R$, then

$$(a,b) \in R \circ R \wedge (b,c) \in R \circ R \Rightarrow \exists x \in A$$

such that

$$(a,x) \in R \wedge (x,b) \in R,$$

and $\exists y \in A$ such that

$$(b,y) \in R \wedge (y,c) \in R.$$

Then

$$(a,b) \in R \wedge (b,y) \in R \Rightarrow (a,y) \in R \circ R$$
$$\Rightarrow (a,y) \in R,$$

and

$$(a,y) \in R \wedge (y,c) \in R \Rightarrow (a,c) \in R \circ R$$
$$\Rightarrow (a,c) \in R.$$

That is,

$$(a,b) \in R \wedge (b,c) \in R \Rightarrow (a,c) \in R.$$

Then R is transitive.

As R is reflexive, antisymmetric, and transitive, then R is a partial order relation on A.

Necessity (see Exercise 1).

Definition 3.6.9 Let A be a partially ordered set by the partial order relation R and let $(a,b) \in A$, such that aRb. We can express that by one of the following diagrams, called Hasse diagram.

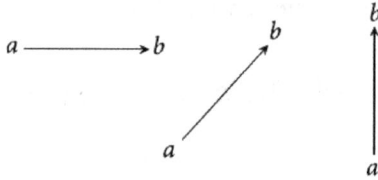

Example 3.6.10 Let $A = \{1,3,5,12\}$ and $R = \{(x,y) \in A \times A : x \le y\}$. It is clear that R is a partial order relation on A. The Hasse diagram of the partial order relation R is

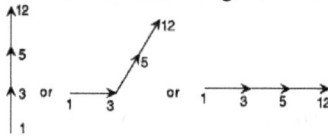

Example 3.6.11 Let $A = \{2,6,20,15\}$ and let $R = \{(x,y) \in A \times A : x \,|\, y\}$. Then R is a partial order relation on A. The Hasse diagram of R is

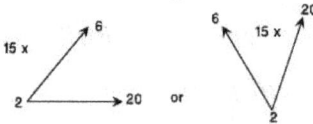

Example 3.6.12 Let $A = \{1,2,3\}$. Let R be a relation on $P(A)$ defined by

$$R = \{A,B) \in P(A) \times P(A) : A \subseteq B\}.$$

R is a partial order relation on $P(A)$. A Hasse diagram for $P(A)$ is

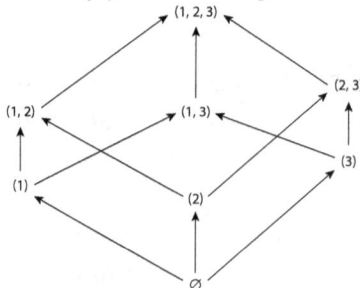

Definition 3.6.13 Let A be a partially ordered set by the relation R. The element $b \in A$ is called the least element of A if and only if

$$bRx, \forall\, x \in A.$$

Example 3.6.14 Let $A = \{3, 6, 9, 12, 15\}$ and let R, S, and T, be relations on A defined by

$$R = \{(x, y) \in A \times A : x \le y\},$$
$$S = \{(x, y) \in A \times A : x \ge y\},$$
$$T = \{(x, y) \in A \times A : x \mid y\}.$$

It is clear that R, S, and T are partial order relations.
The element 3 is the least element for A associated to R, since

$$3 \le x, \forall\, x \in A.$$

The element 15 is the least element for A associated to S, since

$$15 \ge x, \forall\, x \in A.$$

The element 3 is the least element for A associated to T, since

$$3 \mid x, \forall\, x \in A.$$

Example 3.6.15 Let X be any set and let R be a relation on $P(X)$ defined by

$$R = \{(A, B) \in P(X) \times P(X) : A \subseteq B\}.$$

It is clear that R is a partial order relation on $P(X)$.
The empty set ϕ is the least element for $P(X)$ associated to R, since

$$\phi \subseteq A, \forall\, A \in P(X).$$

Theorem 3.6.16 Let R be a partial order relation on a set A. If A has a least element, then it is unique.

Proof Suppose R has two least elements b and c. Then

$$bRc \wedge cRb,$$

by definition of least element. Since R is antisymmetric, then

$$bRc \wedge cRb \Rightarrow b = c.$$

Thus the least element, if exist, is unique.

Definition 3.6.17 Let R be a partial order relation on a set A. The element $a \in A$ is called greatest element of A associated to R if and only if

$$xRa, \forall\, x \in A.$$

Example 3.6.18 Let $A = \{3,5,6,9,10,12,13\}$, and let G,H, and K be relations on A defined by

$$G = \{(x,y) \in A \times A : x \leq y\},$$
$$H = \{(x,y) \in A \times A : x \geq y\},$$
$$K = \{(x,y) \in A \times A : x \mid y\}.$$

The relations G,H, and K are partial order relations on A. The number 13 is the greatest element for A associated to G, since

$$x \leq 13, \forall x \in A.$$

The number 3 is the greatest element for A associated to H, since

$$x \geq 3, \forall x \in A.$$

But there does not exist the greatest element associated to K, because there does not exist an element $a \in A$ that divides by each element of A.

Example 3.6.19 Let X be an arbitrary set, and let R be a relation on $P(X)$ defined by

$$R = \{(A,B) \in P(X) \times P(X) : A \subseteq B\}.$$

R is a partial order relation on $P(X)$. Since $A \subseteq X, \forall\ A \in P(X)$, then X is the greatest element for $P(X)$ associated to R.

Theorem 3.6.20 Let R be a partial order relation on a set A. If A has a greatest element, then it is unique.

Proof Suppose that A has two greatest elements a and b. Then $aRb \wedge bRa$, by definition of the greatest element.
Hence

$$aRb \wedge bRa \Rightarrow a = b,$$

since R is antisymmetric.
Thus the greatest element, if exist, is unique.

Definition 3.6.21 Let R be a partial order relation on a set A. The element $m \in A$ is called maximal element of A associated to R if there does not exist $x \in A$, such that

$$mRx \text{ and } m \neq x.$$

Example 3.6.22 Let $A = \{3,6,7,12,13,15,18\}$ and let R,S, and T be relations on A defined by

$$R = \{(x,y) \in A \times A : x \geq y\},$$
$$S = \{(x,y) \in A \times A : x \leq y\},$$
$$T = \{(x,y) \in A \times A : x \mid y\}.$$

The relations R, S, and T are partial order relations on A. The number 3 is maximal element for A associated to R, since there does not exist $x \in A$ such that

$$3 \geq x \text{ and } 3 \neq x.$$

The number 18 is the maximal element for A associated to S, since there does not exist $x \in A$, such that

$$18 \leq x \text{ and } 18 \neq x.$$

The number 7 is a maximal element for A associated to T, since there does not exist $x \in A$, such that

$$7 \mid x \text{ and } 7 \neq x.$$

Also 12, 13, 15, and 18 are maximal elements associated to T.

Example 3.6.23 Let $X = \{1,3,5,6,7\}$ and $E = \{\{1,3\},\{5,7\},\{1,3,5\},\{1,3,7\}, X\}$. Let G and H be relations on E defined by

$$G = \{(A,B) \in E \times E : A \subseteq B\},$$
$$H = \{(A,B) \in E \times E : A \supseteq B\}.$$

The relations G and H are partial order relations on E. The set X is the maximal element for E associated to G. The set $\{5,7\}$ is a maximal element for E associated to H, since there does not exist $B \in E$, such that

$$\{5,7\} \supseteq B \text{ and } \{5,7\} \neq B.$$

Definition 3.6.24 Let R be a partial order relation on a set A. The element $n \in A$ is called minimal element of A associated to R if there does not exist an element $x \in A$, such that xRn and $x \neq n$.

Example 3.6.25 Let $A = \{3,5,9\}$ and let R, S, and T be relations on A defined by

$$R = \{(x,y) \in A \times A : x \geq y\},$$
$$S = \{(x,y) \in A \times A : x \leq y\},$$
$$T = \{(x,y) \in A \times A : x \mid y\}.$$

R, S, and T are partial order relations on A. The number 3 is the minimal element of A associated to R. The number 9 is the minimal element of A associated to S. Both the numbers 5 and 7 are minimal elements for A associated to T.

Example 3.6.26 Let

$$X = \{a,b,c,d\},$$
$$E = \{\{a,b\},\{b,d\},\{a,b,c\},X\},$$

and let

$$R = \{(A,B) \in E \times E : A \subseteq B\},$$
$$S = \{(A,B) \in E \times E : A \supseteq B\}.$$

R and S are partial order relations on E. Both $\{a,b\}$ and $\{b,d\}$ are minimal elements of E associated to R. X is the minimal element of E associated to S.

Definition 3.6.27 Let A be a partially ordered set with respect to the relation $\leq$, and let B be a subset of A. An element $a \in A$ is called upper bound of B in A if

$$\forall x \in B, a \geq x (xRa, \forall x \in B).$$

In this case, we say that B is bounded above. An element $a \in A$ is called lower bound of B in A if

$$\forall x \in B, a \leq x (aRx, \forall x \in B).$$

In this case, we say that B is bounded below.

Example 3.6.28 Let R be a relation on $\mathbb{R}$ defined by

$$R = \{(x,y) \in \mathbb{R} \times \mathbb{R} : x \leq y\},$$

and let $B = [2,3]$. The number $5 \in \mathbb{R}$ is an upper bound of B. The number $1 \in \mathbb{R}$ is a lower bound of B.

Example 3.6.29 Let $A = \{5,3,10,12,30\}$ and let R be a relation on A defined by

$$R = \{(x,y) \in A \times A : x \mid y\},$$

and let $B = \{5,10\}$. The number $30 \in A$ is an upper bound of B. The number $5 \in A$ is a lower bound of B.

Definition 3.6.30 Let R be a partial order relation on a set A and let B be a subset of A. The element $x \in A$ is called a least upper bound (or supremum) of B if
x is an upper bound of B and
xRy, for every upper bound y of B.
We write $Sup(B)$ (or $LUB(B)$) to denote the supremum of B.

Definition 3.6.31 Let R be a partial order relation on a set A and let B be a subset of A. The element $x \in A$ is called a greatest lower bound (or infimum) of B if
x is a lower bound of B, and
yRx, for every lower bound y of B.
We write $Inf(B)$ (or $GLB(B)$) to denote the infimum of B.

Example 3.6.32 Let T be a relation on R defined by

$$T = \{(x,y) \in \mathbb{R} \times \mathbb{R} : x \leq y\},$$

and let $B = [-1,5]$. It is clear that T is a partial order relation on R. Then $Sup(B) = 5$ and $Inf(B) = -1$.

Remark 3.6.33 Let $\leq$ be a partial order relation on a set A, and let B be a subset of A. Then the following assertion hold.

1. If $\text{Sup}(B)$ exists, it is unique, and if $\text{Inf}(B)$ exists, it is unique.

2. b is an upper bound of B associated to the relation $\leq$ iff b is a lower bound of B associated to the relation $\geq$.

3. $b = \text{Sup}(B)$ associated to $\leq$ iff $b = \text{Inf}(B)$ associated to $\geq$.

Definition 3.6.34 Let R be a partial order relation on a set A. A is said to be a lattice if and only if for any elements $(a,b) \in A$, $Inf\{a,b\}$, and $\text{Sup}\{a,b\}$ exist.

Definition 3.6.35 Let $\leq$ be a partial order relation on a set A. If $x, y \in A$, we say that x, y are comparable if

$$x \leq y \text{ or } y \leq x.$$

Otherwise x, y are incomparable.

Example 3.6.36 Let R be a relation on the set $\mathbb{Z}$ defined by

$$R = \{(x,y) \in \mathbb{Z} \times \mathbb{Z} : x \leq y\}.$$

It is clear that R is a partial order relation on $\mathbb{Z}$. Any pair of numbers $x, y \in \mathbb{Z}$ are comparable because either

$$x \leq y \text{ or } y \leq x.$$

Example 3.6.37 Let R be a relation on $A = \mathbb{Z}^+$ defined by

$$R = \{(x,y) \in A \times A : x \mid y\}.$$

R is a partial order set on A. It is not necessary that any pair $x, y \in A$ are comparable. For example $3, 5 \in A$, but $3 \nmid 5$ and $5 \nmid 3$.

Example 3.6.38 Let $X = \{1, 2, 3\}$ and let R be a relation on $P(X)$ defined by

$$R = \{(A,B) \in P(X) \times P(X) : A \subseteq B\}.$$

R is a partial order relation on $P(X)$. It is not necessary that any pair of subsets A, B are comparable. For example, if $A = \{1\}$, $B = \{5\}$, we see that $A \not\subseteq B$, $B \not\subseteq A$.

Definition 3.6.39 Let R be a partial order relation on a set A, and let B be a subset of A. B is called a totally ordered subset (or B is a chain in A) if any pair of elements $(a,b) \in B$ are comparable, i.e., aRb or bRa.

If any pair of elements $(a,b) \in A$ are comparable, A is called totally ordered set.

Example 3.6.40 The set $\mathbb{Z}$ is a totally ordered set with the relation $\leq$.

Example 3.6.41 Let

$$A = \{1,2,3,4,5,6,7,8,9\},$$
$$B = \{2,4,8\},$$

and

$$R = \{(x,y) \in A \times A : y \mid x\}.$$

The set A is not totally ordered set, but B is a totally ordered set.

Example 3.6.42 Let A, B be totally ordered sets with the relations R and S, respectively. Then the Cartesian product $A \times B$ can be totally ordered by the relation T as follows:

$$(a,c)T(b,d) \text{ if and only if } aRb, \ a,b \in A,$$

or

$$a = b, cSd, c,d \in B.$$

This order is called lexicographical order of $A \times B$. For example, let $A = \{1,3,5\}$, $B = \{2,4\}$ and let

$$R = \{(x,y) \in A \times A : x \le y\},$$
$$S = \{(x,y) \in B \times B : x \mid y\}.$$

It is clear that R and S are total order relations on A and on B, respectively. Suppose T is the lexicographical order of

$$A \times B = \{(1,2),(1,4),(3,4),(3,2),(5,2),(5,4)\}.$$

See that $(1,2)T(1,4)$, since $1 = 1$ and $2 \mid 4$. Also $(1,2)T(3,2)$, since $1 < 3$. The following graph explains $(A \times B, R \times S)$:

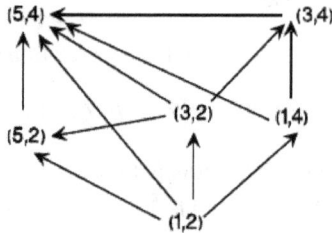

Theorem 3.6.43 Let R be a total order relation on a set A. Then A has at most one minimal element, which is the least element of A. Also A has at most one maximal element, which is the greatest element of A.

Proof Assume that a_1, a_2 are maximal elements with $a_1 \ne a_2$. Then there does not exist $x \in A$, such that

$$xRa_1 \text{ and } x \ne a_1.$$

Also there does not exist $x \in A$, such that

$$xRa_2 \text{ and } x \neq a_2.$$

As A is the totally ordered set, then a_1Ra_2 or a_2Ra_1. This contradicts that $a_1 \neq a_2$. Thus $a_1 = a_2$.

Now, we prove that a_1 is the least element of A, that is $a_1Rx, \forall x \in A$.

Suppose there exists $x \in A$, such that $a_1 \cancel{R} x$. Then xRa_1, since A is the totally ordered set. But this contradicts that a_1 is the least element, thus a_1 is the least element of A.

The proof of the second part is left as Exercise 2.

Definition 3.6.44 Let R be a partial order relation on a set A. Set A is said to be well-ordered set if and only if arbitrary nonempty subset of A has a least element.

Theorem 3.6.45 Every well-ordered set is totally ordered.

Proof Let A be a well-ordered set and $x, y \in A$. Let

$$B = \{x, y\} \subseteq A.$$

Then B has a least element x or y. So, for any two elements $x, y \in A$, we have that x, y are comparable. Thus A is totally ordered.

Example 3.6.46 Let $A = \{2, 3, 4, 5, 6\}$ and $R = \{(x, y) \in A \times A : x \leq y\}$. Thus the set A is well-ordered.

Example 3.6.47 Let $A = \mathbb{N}$ and $R = \{(x, y) \in \mathbb{N} \times \mathbb{N} : x \leq y\}$. Thus $\mathbb{N}$ is well-ordered.

Example 3.6.48 The set of integers $\mathbb{Z}$, with the total order relation $\leq$, is not well-ordered, since the subset

$$A = \{\ldots, -2, -1, 0\} \subset \mathbb{Z}$$

has not a least element.

Theorem 3.6.49 If A is a well-ordered set, then $\forall a \in A$ (except the greatest element) has an immediate successor.

Proof Let
$$T = \{y \in A : y > a\}.$$

T is not empty subset of A, why?

Then T has a least element, say b. Thus b is the immediate successor of a.

Definition 3.6.50 Let R be a partial order relation on a set A, and let $B \subseteq A$. B is called a section of A if

$$\forall x \in A, (y \in B, xRy) \Rightarrow x \in B.$$

Theorem 3.6.51 (Principal of transfinite induction) Let A be a well-ordered set and $P(x)$ is an open sentence on A.
Then

$$[P(y) \text{ is true } \forall \, y < x \Rightarrow P(x) \text{ is true}] \Rightarrow P(x) \text{ is true } \forall \, x \in A.$$

Proof Suppose $P(x)$ is not true for all $x \in A$. We define the set

$$T = \{y \in A : P(x) \text{ is false}\}.$$

As T is a subset of A, then T has a least element, say m. Since $P(x)$ is true $\forall \, x < m$, then $P(m)$ is true. But $P(m)$ is false, since m is the least element of T such that $P(m)$ is false. Thus $P(x)$ is true $\forall \, x \in A$.

Exercises

1. Complete the proof of Theorem 3.6.8.

2. Complete the proof of Theorem 3.6.43.

3. Let $A = \{0, 1, 2\}$. Write all the partial order relations, possible, on A. Which of those relations are total order relations?

4. Is ϕ a partial order relation?

5. Let S be a nonempty family of partial order relations on a set A. Prove that
$$\bigcap_{\delta \in S} \delta \text{ is a partial order relation on } A.$$

6. Let S be a partial order relation on a set X and $A \subseteq X$. Prove that

   (a) $S \cap (A \times A)$ is a partial order relation on A,

   (b) if S is a total order relation on X, then $S \cap (A \times A)$ is a total order relation on A.

7. Let R be a partial order relation on a set X. Prove that $R - I_X$ is a strict order relation on X.

8. Let R be a strict order relation on a set X. Prove that $R \cup I_X$ is a partial order relation on X.

9. Give an example of a set X with a set T of partial order relations on X, such that
$$\bigcup_{\delta \in T} \delta$$
is not a partial order relation.

10. Draw the Hasse diagram for the partial ordered set (X, S), where

$$X = \{a, b, c, d, e\},$$
$$S = \{(a, d), (a, c), (a, b), (a, e), (b, e), (c, e), (d, e)\}.$$

11. Let (X, S) be a partial ordered set, where

$$X = \{a, b, c, d\},$$
$$S = \{(c, d)\} \cup I_X.$$

Determine the maximal elements, minimal elements, the least element, and the greatest element.

12. Prove that every finite set is a well-ordered set.

13. Prove that any subset of a well-ordered set is a well-ordered set.

14. Let $\{S_i\}_{i \in I}$ be a nonempty family of equivalence relations on X, such that

$$(\{S_i\}_{i \in I}, \subseteq)$$

is a total ordered set. Prove that $\bigcup_{i \in I} S_i$ is an equivalence relation on X.

15. Let (A, R), (B, S) be partially ordered sets, and $A \times B$ an ordered set by the lexicographical relation T. Prove that if (a, b) is a greatest element of $A \times B$, then a is a greatest element of A.

16. Show that if R is an antisymmetric relation, then $x \neq y$ and imply $y \not\mathrel{R} x$.

17. Let R be a relation on $\mathbb{N}$ given by

$$R = \{(a, b) \in \mathbb{N} \times \mathbb{N} : b = 2^k \text{ a for some integer } k \geq 0\}.$$

Prove that R is a partial order relation on $\mathbb{N}$.

18. Let $A = \mathbb{R} \times \mathbb{R}$ and let S be a relation on A given by

$$S = \{((a, b), (c, d)) \in A \times A : a \leq c \wedge b \leq d\}.$$

Prove that S is a partial order relation on A.

19. Let T be a relation on C given by

$$T = \{(a + bi, c + di) : a^2 + b^2 \leq c^2 + d^2\}.$$

Is T a partial order relation on C? Justify your answer.

20. Draw the Hasse diagram for the poset $P(X)$ with respect to the relation set inclusion, where

$$A = \{a, b, c, d\}.$$

21. Let X be a set. Consider the partial order relation $\subseteq$ on $P(X)$. Let C and D be subsets of X. Prove that the least upper bound of $\{C,D\}$ is $C \cup D$ and the greatest lower bound of $\{C,D\}$ is $C \cap D$.

22. Let R be a relation on N given by

$$R = \{(m,n) \in \mathbb{N} \times \mathbb{N} : m < 2n\}.$$

Is R a total order relation on $\mathbb{N}$?

23. Let A be a family of subsets of B. Prove that the least upper bound of A is

$$\bigcup_{A \in \mathbf{A}} A,$$

and the greatest lower bound of A is

$$\bigcap_{A \in \mathbf{A}} A.$$

4 Functions

The word "function" was first used by G.W. Leibnitz in 1694. J. Bernoulli defined a function as "any expression including variables and constants" in 1698. The familiar notation $f(x)$ was first used by L. Euler in 1734. We shall present in this chapter the basic properties of functions and induced set functions.

4.1 FUNCTIONS

4.1.1 DOMAIN AND RANGE

Definition 4.1.1 Let A and B be sets and let f be a relation from A to B. f is said to be a function if the following conditions hold:

1. $\forall x \in A$, $\exists y \in B$, such that $(x,y) \in f$.

2. If $(x,y) \in f$ and $(x,z) \in f$, then $y = z$.

By the above definition, a relation f is a function from A to B, if it is wholly defined on A, and it has a unique image in B for an arbitrarily fixed element of A.

Example 4.1.2 Let

$$A = \{1,3,5,7\},$$
$$B = \{2,4,6\},$$

and let f be a relation defined by

$$f = \{(1,2),(3,4),(5,2),(7,6)\}.$$

Then f is a function from A to B.

Example 4.1.3 Let

$$A = \{1,3,5,9\},$$
$$B = \{3,7,11,16,19\},$$

and let f be a relation from A to B defined by

$$f = \{(x,y) : y = 2x+1\}.$$

It is clear that f is a function from A to B.

Example 4.1.4 Let

$$A = \{1,3,5,9\},$$
$$B = \{3,7,11,16\},$$

DOI: 10.1201/9780429022838-4

and let

$$f = \{(x,y) : y = 2x + 1\}.$$

Then f is not a function, since $9 \in A$, but there does not exist $y \in B$, such that $(9, y) \in f$.

A function f from A to B is also called a mapping from A to B.
We write

$$f : A \to B,$$

and this is read "f maps A to B" or "f is a function from A to B". The set A is called the domain of f, denoted by $\mathrm{Dom}(f)$. The set B is called the codomain of f, denoted by $\mathrm{codom}(f)$.

Definition 4.1.5 Let $f : A \to B$. If $(x,y) \in f$, then we write $y = f(x)$, and we say that y is the value of f at x, or the image of x under f. In such a case, x is said to be the pre mage of y under f.

Remark 4.1.6 One can use Venn diagram to clarify a function from A to B.

Example 4.1.7 Let

$$A = \{a, b, c\},$$
$$B = \{2, 1, -1\},$$

We define the function $f : A \to B$ by

$$f = \{(a, 2), (b, 2), (c, 1)\}.$$

The following diagram represents the function f:

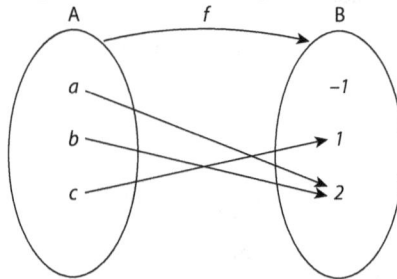

This diagram indeed indicates that the mapping f is wholly defined on A, and each element of A is uniquely assigned to be a single corresponding element of B, and hence, it is a function.

Definition 4.1.8 For every functions $f : A \to B$, we define the range $\mathrm{Rng}(f)$ of f by the relation

$$\mathrm{Rng}(f) = \{y \in B : \exists x \in A, \text{ such that } (x,y) \in f\}$$
$$= \{y \in B : \exists x \in A, \text{ such that } y = f(x)\}.$$

By definition, the range $\text{Rng}(f)$ is the set of all images of A under f. So, sometimes, $\text{Rng}(f)$ is also denoted by $f(A)$.

Remark 4.1.9 If $f : A \to B$ is a function, then

$$\text{Dom}(f) = A \text{ and } \text{Rng}(f) \subseteq B.$$

Example 4.1.10 Let $A = B = \mathbb{R}$ and let

$$f = \{(x,y) : y = x^2\}.$$

Then f is a function from A to B with

$$\text{Rng}(f) = \{y \in \mathbb{R} : y \geq 0\}.$$

Example 4.1.11 Let

$$A = \{0, \frac{1}{4}, \frac{1}{2}, 1, 2, 3\},$$
$$B = \mathbb{R},$$

and let f be a relation from A to B defined by

$$f = \{(x,y) \in A \times B : y = \begin{cases} x^2, & \text{if } x \text{ is integer,} \\ 1/2, & \text{if } x \text{ is not integer.} \end{cases}$$

Then f is a function from A to B and

$$\text{Rng}(f) = \{0, 1, 4, 9, \frac{1}{2}\},$$
$$\text{Dom}(f) = A.$$

4.1.2 GRAPH OF A FUNCTION

In this section, we study the graph of a function $f : A \to B$, as a subset of the Cartesian product set $A \times B$ of A and B.

Definition 4.1.12 Let $f : A \to B$ be a function from A to B. The set

$$G = \{(x,y) \in A \times B : y = f(x)\}$$

is called the graph of the function f.

Remark 4.1.13 If $f : A \to B$ is a function having its graph G, then
 $G \subseteq A \times B$, by the very definition, and
 $G = f$, by regarding f as a relation.

Example 4.1.14 Let $f : \mathbb{R} \to \mathbb{R}$ be a function defined by

$$f(x) = x^2.$$

Then,

$$G = \{(x,y) \in \mathbb{R} \times \mathbb{R} : y = x^2\}$$
$$= \{(0,0),(1,1),(-1,1),(\frac{1}{2},\frac{1}{4}),(-\frac{1}{2},\frac{1}{4}),\ldots\}.$$

Example 4.1.15 Let

$$A = \{1,3,4,5\},$$
$$B = \{3,7,9,11,13,15\},$$

and let $f : A \to B$ be a function defined by

$$f(x) = 2x+1.$$

The graph G of f is

$$G = \{(x,y) \in A \times B : y = 2x+1\}$$
$$= \{(1,3),(3,7),(4,9),(5,11)\}.$$

Notion 4.1.16 Let A and B be sets. The set of all functions from A to B is denoted by

$$B^A.$$

Theorem 4.1.17 If A is a set containing m-many elements and B is a set containing n-many elements, then B^A has n^m-many elements.

Proof Suppose

$$A = \{x_1,x_2,\ldots,x_m\} \text{ and}$$
$$B = \{y_1,y_2,\ldots,y_n\}.$$

Any element of A is related to one element of B in n ways, that is

$$x_1 \text{ is related to } y_1$$
$$\text{or to } y_2$$
$$\vdots$$
$$\vdots$$
$$\vdots$$
$$\text{or to } y_n,$$

$$x \text{ is related to } y_1$$

$$\text{or to } y_2$$

$$\vdots$$

$$\vdots$$

$$\text{or to } y_n.$$

So, every element of $\{x_1, x_2, \ldots, x_m\}$ has n-different ways to be related to one element of $\{y_1, y_2, \ldots, y_n\}$. Thus the number of the ways, making the m-many elements of A related to n-many elements of B, is

$$n \cdot n \cdot n \ldots n(m \text{ times}) = n^m.$$

4.2   ONE-TO-ONE FUNCTIONS AND ONTO FUNCTIONS

In this section, we study certain types of functions, categorized by ways of assignments.

Definition 4.2.1 Let $f : A \to B$ be a function. We say that

"f is a function from A onto B" or "f maps A onto B",

or "f is an onto function,"

if

$$\text{Rng}(f) = B.$$

In other words, "f is onto function" if and only if

$$\forall\, y \in B, \exists x \in A, \text{ such that } y = f(x).$$

An onto function is also called a surjection.

Example 4.2.2 Let $A = B = \mathbb{R}$ and let

$$f = \{(x, y) \in A \times B : y = 5x + 1\}.$$

Then, the function f is onto. Indeed, for all $y \in B$, there exists

$$x = \frac{y - 1}{5}, \text{ where } x \in A,$$

such that

$$f(x) = 5x + 1 = 5 \cdot \frac{y - 1}{5} + 1 = y.$$

Thus $\forall\, y \in B, \exists\, x \in A$, such that $f(x) = y$.

Example 4.2.3 Let $A = \{3,5,7,9,11\}$, $B = \{4,6,8,10\}$, and

$$f = \{(3,4),(5,6),(7,8),(9,10),(11,4)\}.$$

Since $\text{Rng}(f) = \{4,6,8,10\}$, then f is onto function.

Example 4.2.4 Let $A = \{x \in \mathbb{R} : x \geq 0\}$, $B = \mathbb{R}$, and

$$f = \{(x,y) : y = x^2 + 1\}.$$

Since $\text{Rng}(f) = \{y \in B : y \geq 1\} \neq B$, then f is not onto function.

Definition 4.2.5 Let $f : A \to B$ be a function. Then f is called a one-to-one function, if

$$\forall\, a,a' \in A, (a \neq a' \Rightarrow f(a) \neq f(a')).$$

A one-to-one function is also called an injection.

Example 4.2.6 Let $A = \{3,5,8\}$, $B = \{1,2,4,6\}$, and let f,g be functions from A to B defined by

$$f = \{(3,1),(5,4),(8,2)\},$$
$$g = \{(3,2),(5,2),(8,4)\}.$$

Then f is one-to-one. Meanwhile, g is not one-to-one, since

$$3 \neq 5, \text{ but } g(3) = g(5) = 2.$$

Example 4.2.7 Let $A = \{x \in \mathbb{R} : -2 \leq x \leq 5\}$, $B = \mathbb{R}$, and let f,g be functions from A to B defined by

$$f = \{(x,y) \in A \times B : y = x^3\},$$
$$g = \{(x,y) \in A \times B : y = 3x^2 + 1\}.$$

The function f is one-to-one, since

$$x_1^3 = x_2^3 \Rightarrow x_1 = x_2,$$

that is, $f(x_1) = f(x_2) \Rightarrow x_1 = x_2$, which is equivalent to

$$x_1 \neq x_2 \Rightarrow f(x_1) \neq f(x_2).$$

However, the function g is not one-to-one, since

$$-2 = x_1 \neq x_2 = 2; \text{ however }, 13 = f(x_1) = f(x_2) = 13.$$

Example 4.2.8 Let $f : \mathbb{R} \to \mathbb{R}$ be a function defined by

$$f(x) = \frac{1}{x^2 + 1}.$$

Is f one-to-one?

Assume that $f(x_1) = f(x_2)$. Then

$$\frac{1}{x_1^2 + 1} = \frac{1}{x_2^2 + 1}.$$

Therefore, $x_1^2 + 1 = x_2^2 + 1$. So

$$x_1^2 = x_2^2.$$

It does not follow from this that $x_1 = x_2$. For example

$$f(2) = f(-2) = 5,$$

Therefore, f is not one-to-one.

Definition 4.2.9 A function $f : A \to B$, which is both one-to-one and onto is called a bijection, or an one–one correspondence.

Example 4.2.10 Let

$$A = \{1, 3, 5, 7, \ldots\} = \{2n + 1 \mid n \text{ is an integer}\},$$
$$B = \{2, 4, 6, 8 \ldots\} = \{2n \mid n \text{ is an integer}\},$$

and let

$$f = \{(x, y) \in A \times B : y = 2x\},$$
$$g = \{(x, y) \in A \times B : y = x + 1\}.$$

The function $f : A \to B$ is not a bijection, because it is not onto, meanwhile, the function $g : A \to B$ is a bijection, since it is both one-to-one and onto.

Example 4.2.11 Let $A = B = \mathbb{R}$, and let

$$f = \{(x, y) \in \mathbb{R} \times \mathbb{R} : y = 2x^3 - 7\}.$$

Then f is a bijection from A to B.

Definition 4.2.12 Let $f : A \to B$ and $g : C \to D$ be functions. The two functions f and g are equal, denoted by

$$f = g,$$

if

1. $A = C$,

2. $B = D$,

3. $f(x) = g(x), \forall\ x \in A$.

Example 4.2.13 Let $f : \mathbb{R} \to \mathbb{R}$ be a function defined by

$$f(x) = |x|,$$

and let $g : \mathbb{R} \to \mathbb{R}$ be a function defined by

$$g(x) = \sqrt{x^2}.$$

The two functions are equal.

Example 4.2.14 Let $f : \mathbb{R} \to \mathbb{R}$ be a function defined by

$$f(x) = x^2,$$

and let $g : \mathbb{C} \to \mathbb{C}$ be a function defined by

$$g(x) = x^2.$$

The functions are not equal, since $\text{Dom}(f) \neq \text{Dom}(g)$.

Definition 4.2.15 The function $f : A \to A$ is called the identity function, if

$$f(x) = x, \forall x \in A.$$

Example 4.2.16 Let $A = \{1, 2, 3\}$, and let

$$f = \{(x, y) \in A \times A : x = y\}$$
$$= \{(1, 1), (2, 2), (3, 3)\}.$$

Then $f : A \to A$ is the identity function.

Remark 4.2.17 The identity function is one-to-one and onto, and hence, it is a bijection.

Definition 4.2.18 The function $f : A \to B$ is called a constant function, if there exists a fixed element $c \in B$, such that

$$f(x) = c, \forall x \in A.$$

Example 4.2.19 Let $f : \mathbb{R} \to \mathbb{R}$ be defined by

$$f(x) = 3, \forall x \in \mathbb{R}.$$

Then f is a constant function.

Remark 4.2.20 If $f : A \to B$ is a constant function. Then the following assertions are true:

1. If A contains more than one element, then f is not one-to-one.

2. If B contains more than one element, then f is not onto.

Definition 4.2.21 Let A be a nonempty subset of B. Then the function

$$f : A \to B$$

is called an inclusion function (or, an embedding), if

$$f(x) = x, \ \forall \, x \in A.$$

Example 4.2.22 Let $A = \mathbb{N}$, $B = \mathbb{Z}$, and let

$$f = \{(x,y) \in \mathbb{N} \times \mathbb{Z} : y = x\}.$$

Since $\mathbb{N} \subseteq \mathbb{Z}$ and $f(x) = x$, $\forall \, x \in \mathbb{N}$, it follows that $f : \mathbb{N} \to \mathbb{Z}$ is an inclusion function.

Example 4.2.23 Let $A = \{x \in \mathbb{R} : -2 \leq x \leq 2\}$, $B = \mathbb{R}$, and let

$$f = \{(x,y) \in A \times B : y = x\}.$$

Since $A \subset \mathbb{R}$ and $f(x) = x$, then $f : A \to B$ is an inclusion function.

Remark 4.2.24 Let $f : A \to B$ be an inclusion function. Then the following assertions are true:

1. If $A = B$, then $f = I_A$.

2. The function f is one-to-one.

3. If $A \subset B$, then f is not onto.

Definition 4.2.25 Let B be a subset of A and $C = \{0, 1\}$, and let f be the relation from A to C defined by

$$f(x) = \begin{cases} 0 & \text{if } x \in B, \\ 1 & \text{if } x \in A - B. \end{cases}$$

The function $f : A \to C$ is called the characteristic function of B in A.

By definition, characteristic functions are onto $\{0, 1\}$, if and only if B is a non empty proper subset of A. It is not hard to check that a characteristic function of a non empty subset B in A is not onto $\{0, 1\}$, if and only if $B = A$.

Definition 4.2.26 Let $f : A \to B$ be a function, and let C be a subset of A. Then the function,

$$g : C \to B,$$

which is defined by $g(x) = f(x)$, $\forall \, x \in C$, is called the restriction of f to C. We denote this restriction g by

$$f|_C.$$

Example 4.2.27 Let f be a function on $\mathbb{R}$ defined by

$$f = \{(x,y) \in \mathbb{R} \times \mathbb{R} : y = f(x) = x^2\},$$

and let g be a function from $\mathbb{N}$ to $\mathbb{R}$ defined by

$$g = \{(x,y) \in \mathbb{N} \times \mathbb{R} : y = g(x) = x^2\}.$$

Since $\mathbb{N} \subset \mathbb{R}$ and $f(x) = g(x), \forall x \in \mathbb{N}$, then $g : \mathbb{N} \to \mathbb{R}$ is the restriction of f to $\mathbb{N}$.

One can check that if a function is one-to-one, then its restrictions are one-to-one, however, even though a function is onto, the corresponding restrictions are not onto, in general. By definitions, every inclusion function is understood to be a restriction.

Definition 4.2.28 Let $f : A \to B$ be a function, and let $A \subseteq D$. Then the function,

$$g : D \to B,$$

is called an extension of f from A to D, if

$$g(x) = f(x), \forall x \in A,$$

that is $g|_A = f$.

Example 4.2.29 In Example 4.2.27, the function

$$f : \mathbb{R} \to \mathbb{R}$$

is an extension of the function

$$g : \mathbb{N} \to \mathbb{R}.$$

It can be verified that, even though f is onto (or, one-to-one), its extensions are not onto (respectively, one-to-one), in general. That is, the one-to-one-ness, and the onto-ness of functions do not affect those of extensions in general.

Definition 4.2.30 A function $f : A \to B$ is called a numerical function, if B is a set of numbers.

Majority of the exercises and examples above are numerical functions.

Definition 4.2.31 Let $A = B = \mathbb{R}$, and let

$$f = \{(x,y) \in \mathbb{R} \times \mathbb{R} : y = |x|\},$$

where

$$|x| = \begin{cases} x & \text{if } x \geq 0, \\ -x & \text{if } x < 0. \end{cases}$$

This function is called the absolute value function.

By definition, the absolute value function is neither one-to-one nor onto on $\mathbb{R}$.

Definition 4.2.32 Let A be an arbitrary set. A function,

$$f : \mathbb{N} \to A,$$

is called a sequence in A. By denoting the images of f by

$$\{f_n\} \text{ or } f_1, f_2, \ldots, \text{ where } f_n = f(n) \text{ for all } n \in \mathbb{N},$$

one can obtain a sequence in A.

Remark 4.2.33 If $A = \mathbb{R}$, the sequence is called sequence of real numbers.

Example 4.2.34 $\left\{\frac{1}{2^n}\right\}$, $\{(-1)^n\}$ are sequences of real numbers.

Depending on sequences, they can be either one-to-one or onto case-by-case. However, in general, sequences are neither one-to-one nor onto.

Definition 4.2.35 Let A be a nonempty finite set. Every bijection f from A to A is called a permutation of A. That is, all bijections on "finite" sets are said to be permutations.

Example 4.2.36 Let $A = \{1, 3, 5\}$, and let $f : A \to A$ be defined by

$$f(1) = 3,$$
$$f(3) = 5,$$
$$f(5) = 1.$$

Since $f : A \to A$ is a bijection on a finite set A, it is a permutation of A.

Definition 4.2.37 Let A be a set, and R be an equivalence relation on A. The function $f : A \to A/R$, defined by

$$f(x) = [x],$$

is called a canonical function (or, a quotient map).

By the very definition, every canonical function is onto the quotient set A/R, but it is not one-to-one, because if two distinct elements x and y of A are related under an equivalence relation R, then they induce the same equivalence class, that is, $[x] = [y]$, implying $f(x) = f(y)$.

Example 4.2.38 Let $A = \mathbb{Z}$, and let R be the relation on $\mathbb{Z}$ defined by

$$R = \{(x, y) \in \mathbb{Z} \times \mathbb{Z} : y - x \text{ is even}\}.$$

It is clear that R is an equivalence relation on $\mathbb{Z}$ and

$$\mathbb{Z}/R = \{[0], [1]\}.$$

The function $f : \mathbb{Z} \to \mathbb{Z}/R$, defined by $f(n) = [n]$, is a canonical function. For example,

$$f(2) = [2] = [0],$$
$$f(5) = [5] = [1].$$

We remark that $f : \mathbb{Z} \to \mathbb{Z}/R$ is onto but not one–one.

Definition 4.2.39 Let A and B be sets. If f is a function with

$$\mathrm{Dom}(f) = A \times B,$$

then f is called a function of several (or, multi) variables.

Example 4.2.40 Let $f : \mathbb{R} \times \mathbb{R} \to \mathbb{R}$ be a function defined by

$$f(x,y) = \frac{1}{3}\pi x^2 y, \ \forall \ (x,y) \in \mathbb{R} \times \mathbb{R}.$$

Then f is a function of two variables x and y, its domain is $\mathbb{R} \times \mathbb{R}$.

One can generalize this concept with $\mathrm{Dom}(f)$. In addition, $\mathrm{codom}(f)$ are the Cartesian product of more than two sets. For example,

$$f : \mathbb{R}^4 \to \mathbb{R}^2,$$

defined by

$$f(x,y,z,w) = (3x+y, 4z+w),$$

is a function in four variables.

Definition 4.2.41 Let A be a set, and let

$$\mathbb{R}^+ = \{x \in \mathbb{R} : x \geq 0\}.$$

The function $d : A \times A \to \mathbb{R}^+$ is called the distance function (or a metric, or a metric function) on A, if the following conditions hold for $\forall \ a,b,c \in A$:

1. $0 \leq d(a,b)$,

2. $d(a,b) = 0 \Leftrightarrow a = b$,

3. $d(a,b) = d(b,a)$,

4. $d(a,b) \leq d(a,c) + d(c,b)$.

In general, a distance function is neither one-to-one nor onto.

Example 4.2.42 Let d be a relation from $\mathbb{R} \times \mathbb{R}$ to $\mathbb{R}^+$ defined by

$$d(a,b) = |a - b|, \ \forall \ (a,b) \in \mathbb{R} \times \mathbb{R}.$$

Then $d : \mathbb{R} \times \mathbb{R} \to \mathbb{R}^+$ is a distance function. In this case, it is onto.

Example 4.2.43 Let A be a set, and let d be a relation from $A \times A$ to $\mathbb{R}^+$ defined by

$$d(a,b) = \begin{cases} 1 & \text{if } a \neq b, \\ 0 & \text{if } a = b. \end{cases}$$

Then, $d : A \times A \to \mathbb{R}^+$ is a distance function, called the discrete distance (or, the Dirac measure). In this case, if A contains more than one element, then d is onto $\{0,1\}$.

Definition 4.2.44 Let A_1, A_2 be sets. Functions,

$$P_i : A_1 \times A_2 \to A_i \ (i = 1,2),$$

which are defined by

$$P_i(a_1,a_2) = a_i \ (i = 1,2),$$

are called the projection functions (or, projections).

Remark 4.2.45 The function $P_i : A_1 \times A_2 \to A_i$ is onto, but not one-to-one. It is possible to generalize this concept to n sets $A_1, A_2, \ldots, A_n$.
The functions

$$P_i : A_1 \times A_2 \times \ldots A_n \to A_i \ (i = 1,2,\ldots,n)$$

defined by

$$P_i(a_1,a_2,\ldots,a_n) = a_i \ (i = 1,2,\ldots,n),$$

are called the projections of the Cartesian product $A_1 \times A_2 \times \ldots A_n$ on A_i.

Example 4.2.46 The function

$$P_i : \mathbb{R}^2 \to \mathbb{R}, \text{ defined by } P_i(x_1,x_2) = x_i, \ (i = 1,2),$$

is the projection of $\mathbb{R}^2$ on $\mathbb{R}$.

As we have seen, projection functions are onto, but not one-to-one.

Theorem 4.2.47 Let

$$f_1 : B \to A,$$
$$f_2 : C \to A,$$

be functions, such that $B \cap C = \phi$. If $f = f_1 \cup f_2$, then

1. $f : B \cup C \to A$ is a function,

2. $f_1 = f|_B$ and $f_2 = f|_C$.

Proof First we prove that

(a) $(x,y) \in f \wedge x \in B \Leftrightarrow (x,y) \in f_1$,

(b) $(x,y) \in f \wedge x \in C \Leftrightarrow (x,y) \in f_2$.

(a) Suppose that $(x,y) \in f \wedge x \in B$. Then

$$(x,y) \in f \Rightarrow (x,y) \in f_1 \wedge (x,y) \in f_2.$$

If $(x,y) \in f_2$, then $x \in C$, and since $x \in B$, then it follows that $x \in B \cap C$, a contradiction.

Hence, $(x,y) \notin f_2$, and therefore $(x,y) \in f_1$.

Conversely, suppose that $(x,y) \in f_1$. Then $x \in B$.

Since $f = f_1 \cup f_2$, then $(x,y) \in f$. Thus

$$(x,y) \in f_1 \Rightarrow (x,y) \in f \wedge x \in B.$$

So, we get $(x,y) \in f \wedge x \in B \Leftrightarrow (x,y) \in f_1$.

(b) We leave the proof, to the reader, as exercise.

Now, we prove the theorem.

1. Suppose that $x \in B \cup C$, then $x \in B \wedge x \in C$.

    Assume that $x \in B$. Since $f_1 : B \to A$ is a function, then $\exists\, y \in A$, such that $(x,y) \in f_1$.

    Since $f_1 \subseteq f$, then $(x,y) \in f$, that is $\exists\, y \in A$, such that

    $$(x,y) \in f \tag{4.1}$$

    Now, assume that $x \in C$. Since

    $f_2 : C \to A$ is a function, then $\exists\, w \in A$, such that $(x,w) \in f_2$.

    As $f_2 \subseteq f$, then $(x,y) \in f$, that is

    $$\exists w \in A, \text{ such that } (x,w) \in f. \tag{4.2}$$

    Thus, one has

    $\exists\, x \in B \cup C, \exists\, z \in A$, such that $(x,z) \in f$, where $z = y$, or $z = w$.

    Suppose $(x,y_1) \in f \wedge (x,y_2) \in f$. Then $x \in B \cup C$ by definition of the function. Thus $x \in B \vee x \in C$. Assume $x \in B$, from (a), that we find

    $$(x,y_1) \in f \wedge x \in B \Rightarrow (x,y_1) \in f_1$$

    also,

    $$(x,y_2) \in f \wedge x \in B \Rightarrow (x,y_2) \in f_1.$$

    Since $f_1 : B \to A$ is a function, then $y_1 = y_2$.

    Now, assume $x \in C$. From (b), we have

    $$(x,y_1) \in f \wedge x \in C \Rightarrow (x,y_1) \in f_2,$$
    $$(x,y_2) \in f \wedge x \in C \Rightarrow (x,y_2) \in f_2.$$

    Since $f_2 : C \to A$ is a function, then $y_1 = y_2$.

    Thus in both cases, we have $y_1 = y_2$. Hence $f : B \cup C \to A$ is a function.

2. We leave the proof of 2, as an exercise for readers.

Exercises

1. Let $f : [1, \infty) \to \mathbb{R}$ be a function defined by

$$f(x) = 4x - 1.$$

Find $\mathrm{Dom}(f)$.

2. Draw the graph of the following relations and determine which of the following graphs represents a function:

   (a) $f = \{(x, y) \in \mathbb{R} \times \mathbb{R} : 2x - y = 4\}$.

   (b) $g = \{(x, g(x)) \in \mathbb{R} \times \mathbb{R} : g(x) = x^2 + 4\}$.

   (c) $h = \{(x, y) : y = \begin{cases} 2x, & x \in (-2, 4). \\ \frac{3x+1}{2}, & x \in (-4, 2). \end{cases}$

3. Discus each of the following statements:

   (a) the volume of a sphere is a function of its radius,

   (b) the radius of a sphere is a function of its volume.

4. Let by $f(n)$ be denoted the number of prime numbers. Find $f(5)$, $f(79)$, if $f(n) \leq n$.

5. Let $f : A \to B$ be one-to-one function, and let $C \subseteq A$.

   Prove that $f|_C : C \to B$ is a one-to-one function.

6. Let $f : A \to B$, $g : C \to D$ be functions. We define

$$(f \times g)(x, y) = (f(x), g(y)), \forall (x, y) \in A \times C.$$

   Prove that

   (a) $f \times g$ is a function from $A \times C$ to $B \times D$,

   (b) if f and g are one-to-one (onto) functions, then $f \times g$ is a one-to-one (onto) function.

7. Let $f : B \to A$, $g : C \to A$ be functions and assume that

$$f|_{B \cap C} = g|_{B \cap C}.$$

   If $h = f \cup g$, prove that

$$h : B \cup C \to A$$

   is a function with $g = h|_C$ and $f = h|_B$.

8. Let S and T be nonempty sets. Prove that there exists a bijection between $S \times T$ and $T \times S$.

9. Let A and B be sets. Prove that

$$B^A = \phi \Leftrightarrow A = \phi \wedge B = \phi.$$

10. Let A be a set, and let

$$f : A \to A$$

be a function defined by $f(x) = x$, $\forall\, x \in A$. Prove that f is a bijection.

11. Let $f : A \to B$, $g : A \to B$ be functions. Prove that if $f \subseteq g$, then $f = g$.

12. Prove that each of the following is a Metric for the indicated set:

(a) $X = \mathbb{R} \times \mathbb{R}$, $d((x,y),(z,w)) = \sqrt{(x-z)^2 + (y-w)^2}$,

(b) $X = \mathbb{R} \times \mathbb{R}$, $d((x,y),(z,w)) = |x-z| + |y-w|$.

13. Suppose A has m elements and B has n elements. We have seen that $A \times B$ has mn elements and that there are 2^{mn} relations from A to B. Find the number of relations from A to B that are

(a) functions from A to B,

(b) functions with one element in the domain,

(c) functions two elements in the domain,

(d) function with whose domain is a subset of A.

14. For the canonical map $f : \mathbb{Z} \to \mathbb{Z}_6$, find

(a) $f(3)$,

(b) The image of 6,

(c) All the pre images of [3],

(d) All the pre images of [1].

15. Explain why the functions $f(x) = \frac{9-x^2}{x+3}$ and $g(x) = 3 - x$ are not equal.

16. Show that the following relations are not functions on $\mathbb{R}$:

(a) $\{(x,y) \in \mathbb{R} \times \mathbb{R} : x^2 = y^2\}$,

(b) $\{(x,y) \in \mathbb{R} \times \mathbb{R} : x^2 + y^2 = 1\}$,

(c) $\{(x,y) \in \mathbb{R} \times \mathbb{R} : x = \cos y\}$,

(d) $\{(x,y) \in \mathbb{R} \times \mathbb{R} : y = \sqrt{x}\}$.

17. Let G be a relation from A to B. Prove that G is the graph of the function

$$f : A \to B$$

if and only if, there exist relations J and H from A to B, such that

$$(H \cap J) \circ G = (H \circ G) \cap (J \circ G).$$

18. Let $f : \mathbb{Z} \times \mathbb{Z} \to \mathbb{Z}$ be a function defined by

$$f(x,y) = xy.$$

Show that f is onto.

19. Let $f : \mathbb{N} \times \mathbb{N} \to \mathbb{N}$ be a function given by

$$f(m,n) = 2^{m-1}(2n-1).$$

Show that f is onto.

20. Let $f : \mathbb{N} \times \mathbb{N} \to \mathbb{N}$ be a function given by

$$f(m,n) = 2^{m-1}(2n-1).$$

Show that f is one-one.

21. Let $f : [0,\infty) \to [0,\infty)$ be a function defined by

$$f(x) = x^2.$$

Show that f is one-one.

22. For each pair of functions h and g, determine whether $h \cup g$ is a function.

(a) $h : (-\infty, 0] \to \mathbb{R}$
$h(x) = 3x + 4$
$g : (0, \infty) \to \mathbb{R}$
$g(x) = \frac{1}{x}$.

(b) $h : [-1, \infty) \to \mathbb{R}$
$h(x) = x^2 + 1$
$g : (-\infty, -1] \to \mathbb{R}$
$g(x) = x + 3$.

(c) $h : (-\infty, 1] \to \mathbb{R}$
$h(x) = |x|$
$g : [0, \infty) \to R$
$g(x) = 3 - |x - 3|$.

(d) Let $f_1 : \mathbb{R} \to \mathbb{R}$ and $f_2 : \mathbb{R} \to \mathbb{R}$. Define the point-wise sum $f_1 + f_2$ and point-wise product $f_1.f_2$ as follows:

$$f_1 + f_2 = \{(a, c+d) : (a,c) \in f_1 \wedge (a,d) \in f_2\},$$
$$f_1.f_2 = \{(a, cd) : (a,c) \in f_1 \wedge (a,d) \in f_2\}.$$

(i) Prove that $f_1 + f_2$ and $f_1.f_2$ are functions with domain $\mathbb{R}$.
(ii) Show that $(f_1 + f_2)(x) = f_1(x) + f_2(x)$ and that

$$(f_1.f_2)(x) = f_1(x).f_2(x).$$

23. Complete the proof of Theorem 4.2.47.

4.3   COMPOSITE FUNCTIONS AND INVERSE FUNCTIONS

In this section, we consider a certain transitivity on functions, called the composition. Note that such a property holds well since every function is defined wholly from its domain, and assigns each element of its domain to a unique image in its range.

Theorem 4.3.1 Let $f : A \to B$ and $g : B \to C$ be functions. Then

$$g \circ f : A \to C \text{ is a function.}$$

Proof Let $x \in A$. Since $f : A \to B$ is a function, then $\exists y \in B$, such that $(x,y) \in f$. Similarly, since $g : B \to C$ is a function, then $\exists z \in C$, such that $(y,z) \in g$. Then $(x,z) \in g \circ f$, by definition of composite relations. That is, $\forall x \in A$, $\exists z \in C$, such that

$$(x,z) \in g \circ f. \tag{4.3}$$

Now, suppose that

$$(x,z_1) \in g \circ f \wedge (x,z_2) \in g \circ f.$$

Then

$$(x,z_1) \in g \circ f \Rightarrow \exists y_1 \in B,$$

such that

$$(x,y_1) \in f \wedge (y_1,z_1) \in g$$

and

$$(x,z_2) \in g \circ f \Rightarrow \exists y_2 \in B,$$

such that

$$(x,y_2) \in f \wedge (y_2,z_2) \in g.$$

As $f : A \to B$ is a function, then

$$(x,y_1) \in f \wedge (x,y_2) \in f \Rightarrow y_1 = y_2.$$

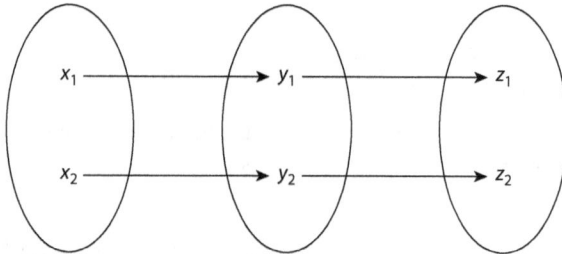

Thus $(y_1,z_2) \in g$.
Since $g : B \to C$ is a function, then

$$(y_1,z_1) \in g \wedge (y_1,z_2) \in g \Rightarrow z_1 = z_2,$$

so, if

$$(x,z_1) \in g \circ f \wedge (x,z_2) \in g \circ f,$$

then

$$z_1 = z_2. \tag{4.4}$$

By Equations (4.3) and (4.4), one can conclude that

$$g \circ f : A \to C$$

is a function.

The above theorem can be re-stated as follows:

Corollary 4.3.2 Let $f : A \to B$, $g : B \to C$ be functions. Then

$$\forall x \in A, (g \circ f)(x) = g(f(x)).$$

Proof Suppose that $z = (g \circ f)(x)$. Then $(x, z) \in g \circ f$. Therefore $\exists\, y \in B$, such that $(x, y) \in f \wedge (y, z) \in g$, but

$$(x, y) \in f \Leftrightarrow y = f(x) \text{ and } (y, z) \in g \Leftrightarrow z = g(y).$$

Hence $z = g(y) = g(f(x))$, so $(g \circ f)(x) = g(f(x))$.

Definition 4.3.3 Let $f : A \to B$, $g : B \to C$ be functions. The function $g \circ f : A \to C$, which is defined by $g \circ f := g(f(x))$, $\forall x \in A$, is called the composite function of f and g. And the construction of composite functions, or the operation ($\circ$) is called the composition.

One can illustrate the composition diagramatically as follows:

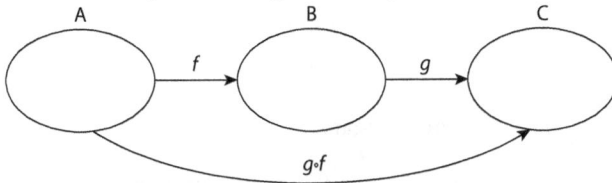

Example 4.3.4 Let $f : A \to B$, $g : B \to C$ be functions defined by the diagram,

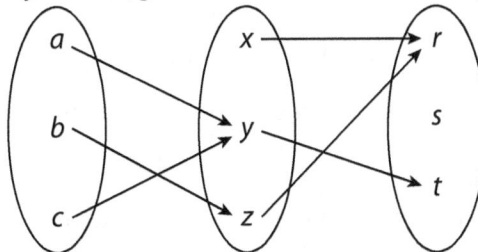

Then the composite function $g \circ f : A \to C$ is defined to be a function satisfying,

$$(g \circ f)(a) = g(f(a)) = g(y) = t,$$
$$(g \circ f)(b) = g(f(b)) = g(z) = r,$$
$$(g \circ f)(c) = g(f(c)) = g(y) = t,$$

having its domain A, and its range contained in C. That is, it satisfies the following diagram:

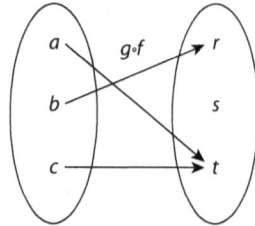

Example 4.3.5 Let $f : \mathbb{R} \to \mathbb{R}$ be the function defined by $f(x) = 4x^2$, and $g : \mathbb{R} \to \mathbb{R}$ be the function defined by $g(x) = 3x + 1$. Then $g \circ f : \mathbb{R} \to \mathbb{R}$ is the function defined by

$$(g \circ f)(x) = g(f(x))$$
$$= g(4x^2)$$
$$= 3(4x^2) + 1$$
$$= 12x^2 + 1,$$

and $f \circ g : \mathbb{R} \to \mathbb{R}$ is the function defined by

$$(f \circ g)(x) = f(g(x))$$
$$= f(3x + 1)$$
$$= 4(3x + 1)^2$$
$$= 36x^2 + 24x + 4.$$

Note that $f \circ g$ and $g \circ f$ are not the same functions.

The above example demonstrates that the composite functions $f \circ g$ and $g \circ f$ are not identical, in general, and hence, the composition ($\circ$) is not commutative (as an operation on functions).

Theorem 4.3.6 Let $f : A \to B$, $g : B \to C$ be functions.

1. If f and g are one-to-one, then $g \circ f : A \to C$ is one-to-one.

2. If f and g are onto, then $g \circ f : A \to C$ is onto.

3. If f and g are bijective, then $g \circ f : A \to C$ is bijective.

Proof By Theorem 4.3.1, the composite function $f \circ g$ is indeed a well-defined function. So, it suffices to check the one-to-one-ness, onto-ness, and bijectivity, respectively.

1. Suppose $x_1, x_2 \in A$, such that $(g \circ f)(x_1) = (g \circ f)(x_2)$.

Then
$$g(f(x_1)) = g(f(x_2)).$$

Since g is one-to-one,
$$g(f(x_1)) = g(f(x_2)) \Rightarrow f(x_1) = f(x_2),$$

and since f is one-to-one,
$$f(x_1) = f(x_2) \Rightarrow x_1 = x_2.$$

Thus, the composition function $g \circ f : A \to C$ is an one–one function.

2. Let $z \in C$. Since g is an onto function,
$$\exists\, y \in B, \text{ such that } z = g(y),$$

and since f is onto,
$$\exists\, x \in A, \text{ such that } y = f(x).$$

Therefore, $\exists\, x \in A$, such that $z = g(f(x)) = (g \circ f)(x)$.

Thus, the composite function $g \circ f : A \to C$ is onto.

3. Since the composite function $g \circ f$ is one-to-one whenever f and g are, and since $g \circ f$ is onto whenever f and g are, it is not hard to verify that $g \circ f$ is bijective, if f and g are bijective. We leave the detailed proof to the readers.

Now, let's consider the converse of Theorem 4.3.6. The converse does not hold true in general, however, we obtain the following results.

Theorem 4.3.7 Let $f : A \to B$ and $g : B \to C$ be function.

1. If $g \circ f : A \to C$ is one-to-one, then f is one-to-one,

2. If $g \circ f : A \to C$ is onto, then g is onto.

Proof By definition, if the composition $g \circ f$ is well-defined as a function, then both f and g are well-defined functions.

1. Let $x_1, x_2 \in A$, such that
$$f(x_1) = f(x_2).$$

Then $g(f(x_1)) = g(f(x_2))$. So $(g \circ f)(x_1) = (g \circ f)(x_2)$. As $g \circ f$ is one-to-one, then it follows that
$$(g \circ f)(x_1) = (g \circ f)(x_2) \Rightarrow x_1 = x_2.$$

Thus f is one-to-one.

2. Let $z \in C$. Since $g \circ f$ is onto, then $\exists\, x \in A$ such that

$$(g \circ f)(x) = z,$$

that is, $\exists\, x \in A$, such that $g(f(x)) = z$. Moreover, $f(x) \in B$. Hence g is onto.

By the above theorem, we obtain the following result, too.

Theorem 4.3.8 Let $f : A \to B$ and $g : B \to C$ be functions. If $g \circ f$ is a bijection, then f is one-to-one and g is onto.

We leave the proof of Theorem 4.3.8 to the readers.

Remark 4.3.9 Note that the converse of Theorem 4.3.8 does not hold true. Even though f is one-to-one, and g is onto, the composite function $g \circ f$ is not necessarily a bijection. See Example 4.3.10 below.

Example 4.3.10 Let $f : \mathbb{R} \to \mathbb{R}$ be the function defined by

$$f(x) = x,$$

and let $g : \mathbb{R} \to \mathbb{R}^+$ be the function defined by

$$g(x) = x^2.$$

Then $g \circ f : \mathbb{R} \to \mathbb{R}^+$ is defined by

$$(g \circ f)(x) = x^2.$$

We see that f is one-one and g is onto, but $g \circ f$ is not bijective.

Definition 4.3.11 If $f : A \to B$ is a function, then the inverse of f is the "relation",

$$f^{-1} = \{(x,y) : (y,x) \in f\}.$$

Remark 4.3.12 By Definition 4.3.11, it is not necessary that f^{-1} is a function from B to A, in general. Also if

$$f^{-1} : B \to A$$

is a function, it is not necessary that $f : A \to B$ is a function.

Example 4.3.13 Let

$$A = \{5,7,9\},$$
$$B = \{-2,4\},$$

and let $f = \{(5,-2),(7,-2),(9,4)\}$. Then the corresponding inverse f^{-1} is determined to be

$$f^{-1} = \{(-2,5),(-2,7),(4,9)\},$$

as a relation. It is clear that $f : A \to B$ is a function, but $f^{-1} : B \to A$ is not a function.

Example 4.3.14 Let $A = \{3\}$, $B = \{2,4,6\}$, and let

$$f = \{(3,2),(3,4),(3,6)\},$$
$$f^{-1} = \{(2,3),(4,3),(6,3)\}.$$

It is clear that $f : A \to B$ is a relation, which is not a function, but $f^{-1} : B \to A$ is a function.

Definition 4.3.15 A function $f : A \to B$ is said to be invertible, if the inverse $f^{-1} : B \to A$ is a function. In this case, the function f^{-1} is called the inverse function of f.

Remark 4.3.16 If $f : A \to B$ is a function, then

$$(x,y) \in f \text{ if and only if } (y,x) \in f^{-1}.$$

In other words, $y = f(x)$, if and only if $x = f^{-1}(y)$. Therefore, $(f^{-1})^{-1} = f$.

Example 4.3.17 Let $f : \mathbb{R} \to \mathbb{R}$ be a function, such that

$$f = \{(x,y) \in \mathbb{R} \times \mathbb{R} : y = x^3\} \Rightarrow f^{-1} = \{(x,y) \in \mathbb{R} \times \mathbb{R} : x = y^3\}.$$

It is clear that $f^{-1} : \mathbb{R} \to \mathbb{R}$ is a function. And hence, f is invertible with its inverse function f^{-1}.

Example 4.3.18 Let $g : \mathbb{R} \to \mathbb{R}$ be a function defined by

$$g = \{(x,y) : y = x^2\}.$$

Then,

$$g^{-1} = \{(x,y) \in \mathbb{R} \times \mathbb{R} : x = y^2\}.$$

So, the relation g^{-1} is not a function, equivalently, the function $f : \mathbb{R} \to \mathbb{R}$ is not invertible.

The following theorem gives us a characterization for a function to be invertible.

Theorem 4.3.19 The function $f : A \to B$ is invertible, if and only if it is bijective.

Proof Suppose $f : A \to B$ is invertible, then the inverse $f^{-1} : B \to A$ of f is a function. So, it suffices to prove that $f : A \to B$ is both one-to-one and onto.
Suppose $x_1, x_2 \in A$, such that $f(x_1) = f(x_2) = y$.
Then,

$$(x_1,y) \in f \wedge (x_2,y) \in f \Rightarrow (y,x_1) \in f^{-1} \wedge (y,x_2) \in f^{-1}$$
$$\Rightarrow x_1 = x_2.$$

since the inverse f^{-1} is a function. Thus $f : A \to B$ is one-to-one.
Now, let $y \in B$. Since $f^{-1} : B \to A$ is a function, then $\exists x \in A$, such that $(y,x) \in f^{-1}$. So, $\exists x \in A$, such that $(x,y) \in f$; that is, $\exists x \in A$, such that $y = f(x)$. Thus $f : A \to B$ is onto.

Therefore, $f : A \rightarrow B$ is a bijective function.

Conversely, suppose $f : A \rightarrow B$ is a bijective function. Show that $f : A \rightarrow B$ is invertible, that is, the inverse $f^{-1} : A \rightarrow B$ is a function.

First, we assume $y \in B$. Since $f : A \rightarrow B$ is onto, then $\exists x \in A$ such that $f(x) = y$; that is, $\exists x \in A$, such that $(x,y) \in f$. But

$$(x,y) \in f \Rightarrow (y,x) \in f^{-1}.$$

Thus $\exists x \in A$, such that $(y,x) \in f^{-1}$; i.e.,

$$\mathrm{Dom}(f^{-1}) = B \tag{4.5}$$

Now, suppose $(y,x_1) \in f^{-1} \wedge (y,x_2) \in f^{-1}$. Then,

$$(x_1,y) \in f \wedge (x_2,y) \in f; \ i.e., \ f(x_1) = y \wedge f(x_2) = y,$$

implying that, $f(x_1) = f(x_2)$.

Since f is one-to-one, then $f(x_1) = f(x_2) \Rightarrow x_1 = x_2$. Thus $(y,x_1) \in f^{-1} \wedge (y,x_2) \in f^{-1} \Rightarrow$

$$x_1 = x_2. \tag{4.6}$$

Therefore, from Equations (4.5) and (4.6), the inverse

$$f^{-1} : B \rightarrow A$$

is a function.

By Theorem 4.3.19 and by the invertibility $(f^{-1})^{-1} = f$, we obtain the following result.

Corollary 4.3.20 If $f : A \rightarrow B$ is an invertible function, then

$$f^{-1} : B \rightarrow A$$

is a bijective function.

Proof Suppose f is invertible, then the inverse f^{-1} of f is a function.

1. Let $y_1, y_2 \in B$, such that

$$f^{-1}(y_1) = f^{-1}(y_2) = x.$$

Then

$$(y_1,x) \in f^{-1} \wedge (y_2,x) \in f^{-1}.$$

So,

$$(x,y_1) \in f \wedge (x,y_2) \in f.$$

That is

$$y_1 = f(x) \wedge y_2 = f(x).$$

This implies $y_1 = y_2$. Thus, $f^{-1} : B \rightarrow A$ is one-to-one.

2. Let $x \in A$. Since $f : A \to B$ is a function, then

$$\exists \, y \in B \text{ such that } (x,y) \in f,$$

showing that $\exists \, y \in B$, such that $(y,x) \in f^{-1}$. In other words, $\exists \, y \in B$, such that $f^{-1}(y) = x$. Thus, $f^{-1} : B \to A$ is onto.

Therefore, the inverse function $f^{-1} : B \to A$ of f is a bijection.

Thus, one can find connections between the composition ($\circ$) and the invertibility.

Theorem 4.3.21 If $f : A \to B$ is invertible function, then

1. $f^{-1} \circ f = I_A$,

2. $f \circ f^{-1} = I_B$,

where I_A and I_B are the identity functions on A, respectively, on B.

Proof We here prove part 1, and leave the proof of part 2, for readers, as exercise.

1. Since $f : A \to B$ is invertible, then

$$f^{-1} : B \to A$$

is a function, and $f^{-1} \circ f : A \to A$ is a function. Suppose $x \in A$ and $f(x) = y$. Then

$$(f^{-1} \circ f)(x) = f^{-1}(f(x)) = f^{-1}(y) = x.$$

Since $I_A : A \to A$ is defined by

$$I_A(x) = x, \ \forall \, x \in A,$$

i.e., $\forall \, x \in A$, $(f^{-1} \circ f)(x) = x = I_A(x)$. i.e., $f^{-1} \circ f = I_A$.

Exercises

1. Let $f : A \to B$, $g : B \to C$, and $h : C \to D$ be functions. Then $h \circ (g \circ f) : A \to C$ is a function. Prove that

$$h \circ (g \circ f) = (h \circ g) \circ f.$$

2. Find $f \circ g$ and $g \circ f$ for each pair of functions f and g.

   (a) $f : \mathbb{R} \to \mathbb{R}$ given by $f(x) = 2x + 5$,
   $g : \mathbb{R} \to \mathbb{R}$ given by $g(x) = 6 - 7x$.

   (b) $f : \mathbb{R} \to \mathbb{R}$ given by $f(x) = x^2 + 2x$,
   $g : \mathbb{R} \to \mathbb{R}$ given by $g(x) = 2x + 1$.

(c) $f : \mathbb{R} \to \mathbb{R}$ given by $f(x) = \frac{x+1}{x+2}$,

$g : \mathbb{R} \to \mathbb{R}$ given by $g(x) = x^2 + 1$.

3. Find the domain and range of each composite function in Exercise 2.

4. Describe two extensions with domain $\mathbb{R}$ for the function

(a) $f = \{(x,y) \in \mathbb{N} \times \mathbb{N} : y = x^2\}$,

(b) $f = \{(x,y) \in \mathbb{N} \times \mathbb{N} : y = 3\}$.

5. Let $A = \mathbb{R} - \{3\}$ and $B = \mathbb{R} - \{1\}$. Let the function

$$f : A \to B$$

be defined by $f(x) = \frac{x-2}{x-3}$. Show that f is both one-one and onto.

Find a formula for f^{-1}.

6. Suppose $f : A \to B$. Which of the following assertions is always true?

(a) $f(A) \subset B$.

(b) $f(A) = B$.

(c) $f(A) \supset B$.

[a.]

7. Let $f : A \to B$, $g : B \to A$, and let $g \circ f = I_A$. Are the following assertions true or false?

(a) $g = f^{-1}$.

(b) f is an one-one function.

(c) f is onto function.

(d) g is an onto function.

(e) g is an one-to-one function.

8. Consider the following functions:

(a) $f_1 : [-2,2] \to \mathbb{R}$,

(b) $f_2 : [0,3] \to \mathbb{R}$,

(c) $f_3 : [-3,0] \to \mathbb{R}$,

(d) $f_4 : [-5,3) \to \mathbb{R}$.

Find the rang of f_1, f_2, f_3, and f_4, if each of these functions is defined by the same formula

$$f_i(x) = x^2, \; i = 1,2,3,4.$$

9. Let $f : X \to Y$ and $g : Y \to X$ be functions, and let $g \circ f = I_X$.

   Prove that f is one-to-one and g is onto.

10. Let $f : X \to Y$ and $g : Y \to X$ be functions. Let $g \circ f = I_X$ and $f \circ g = I_Y$. Prove that f and g are bijective functions and $g = f^{-1}$.

11. Let $h \in X^X$. Prove that h is one-one if and only if

$$h \circ f = h \circ g \Rightarrow f = g, \ \forall \, f, g \in X^X.$$

12. Let $h \in X^X$. Prove that h is onto, if and only if

$$f \circ h = g \circ h \Rightarrow f = g, \ \forall \, f, g \in X^X.$$

13. Let $f : A \to B$ be a function and let $C \subseteq A$. Prove that

$$f|_C = f \circ E_C,$$

   where E_C is the inclusion function from C to A.

14. Let $h : B \to C$ and $g : B \to C$ be functions. Suppose $g \circ f = h \circ f$ for each function $f : A \to B$. Prove that $g = h$.

15. Let $h : A \to B$ and $g : A \to B$ be functions, and let C be a set, which contains more than one element. Suppose $f \circ g = f \circ h$, for each function $f : B \to C$. Prove that $g = h$.

16. Prove that a function $f : A \to B$ is one-one, if and only if there exists a function $g : B \to A$, such that $g \circ f = I_A$.

17. Let $f : A \to B$ be a function. Prove that

$$I_B \circ f = f \text{ and } f \circ I_A = f.$$

18. Prove that $f : A \to B$ is invertible, if and only if there exists a function $g : B \to A$, such that $f \circ g = I_B$ and $g \circ f = I_A$.

19. Let X and Y be sets. Prove that $X^Y = Y^X \Rightarrow X = Y$.

20. Let A, B, and W be sets. Prove that

$$A \subseteq B \Rightarrow A^W \subseteq B^W.$$

21. Let A, B be sets. The symbol $A \sim B$ denotes that there exists a bijective function from A to B. Prove that

$$A \sim B \Rightarrow A^W \sim B^W, \text{ for any set } W.$$

22. Let A, B, and C be sets, such that $B \cap C = \phi$.

   (a) Prove that $A^{B \cup C} \sim A^B \times A^C$.

(b) Is $(A^B)^C \sim A^{B \times C}$?

(c) Prove that $(A \times B)^C \sim A^C \times B^C$.

23. Give an example of functions $f : A \to B$ and $g : B \to C$, such that

(a) f is onto B, but $g \circ f$ is not onto C,

(b) g is onto C, but $g \circ f$ is not onto C,

(c) f is one-to-one, but $g \circ f$ is not one-one,

(d) $g \circ f$ is onto C, but f is not onto B,

(e) g is one-to-one, but $g \circ f$ is not one-to-one,

(f) $g \circ f$ is one-to-one, but g is not one-to-one.

24. Which of the following functions maps onto its indicated codomain? Justify your answers:

(a) $f : \mathbb{R} \to \mathbb{R}$
$f(x) = \frac{1}{2}x^2 + 6$

(b) $f : \mathbb{R} \to \mathbb{R}$
$f(x) = x^2$

(c) $f : \mathbb{R} \to \mathbb{R}$
$f(x) = 2^x$

(d) $f : \mathbb{R} \to [-1, 1]$
$f(x) = \cos x$

(e) $f : (1, \infty) \to (1, \infty)$
$f(x) = \frac{x}{x-1}$

(f) $f : [2, 3] \to [0, \infty]$
$f(x) = \frac{x-2}{3-x}$

(g) $f : \mathbb{R} \to [1, \infty)$
$f(x) = x^2 + 1$

(h) $f : \mathbb{R} \to \mathbb{R}$
$f(x) = x^3$

(i) $f : \mathbb{R} \to \mathbb{R}$
$f(x) = \sin x$

(j) $f : \mathbb{R} \times \mathbb{R} \to \mathbb{R}$
$f(x, y) = x - y$

(k) $f : \mathbb{R} \to \mathbb{R}$
$f(x) = \sqrt{x^2 + 5}$

25. Which of the functions in Exercise 24 are one-to-one? Justify your answers.

26. Prove that

(a) $f(x) = \begin{cases} \frac{x-2}{x-4}, & \text{if } x \neq 4, \\ 1, & \text{if } x = 4 \end{cases}$

is one-to-one and onto $\mathbb{R}$,

(b) $f(x) = \begin{cases} x+4, & \text{if } x \leq -2, \\ -x, & \text{if } -2 < x < 2 - x, \\ x - 4 & \text{if } x \geq 2 \end{cases}$

is onto $\mathbb{R}$, but not one-to-one.

27. Prove that

(a) $f : \mathbb{Z}_4 \to \mathbb{Z}_8$ defined by

$$f([x]) = 2[x]$$

is one-one, but not onto,

(b) $f : \mathbb{Z}_4 \to \mathbb{Z}_2$ defined by

$$f([x]) = [3x]$$

is one-one but not onto,

(c) $f : \mathbb{Z}_6 \to \mathbb{Z}_6$ defined by

$$f([x]) = [x+1]$$

is one-one and onto,

(d) $f : \mathbb{Z}_4 \to \mathbb{Z}_4$ defined by

$$f([x]) = [2x]$$

is neither one-to-one nor onto.

28. Suppose the set A has n elements and the set B has m elements. Find the number of

(a) one–one functions from A to B, assuming that

1. $m < n$ 2. $m = n$ 3. $m > n$

(b) onto functions from A to B, assuming that

1. $m = n$ 2. $m > n$ 3. $m = n+1$

4.4   DIRECT IMAGES AND INVERSE IMAGES UNDER A FUNCTION

Definition 4.4.1 Let $f : A \to B$ be a function and let $C \subseteq A$. The set

$$f(C) = \{y \in B : y = f(x) \text{ for some } x \in C\}$$

is called the direct image of C under f.

By definition, if $f : A \to B$ is a function, then the direct image $f(A)$ of A under f is nothing but the range $\text{Rng}(f)$ of f. Also, direct images of all subsets of A become subsets of B, under an additional axiomatization: the direct image of the empty set becomes the empty set.

Example 4.4.2 Let $f : \mathbb{Z} \to \mathbb{R}$ be a function defined by

$$f(x) = x^2 + 2.$$

Let $C = \{-3, -1, 0, 1, 3\}$, then $f(C) = \{11, 3, 2\}$.

Example 4.4.3 Let $f : \mathbb{R} - \{0\} \to \mathbb{R}$ be a function defined by

$$f(x) = \frac{1}{x}.$$

Let $C = (0, 1]$, then $f(C) = \{y \in R : 1 \leq y < \infty\}$.

Theorem 4.4.4 Let $f : X \to Y$ be a function and let A, B be subsets of X. If $A = B$, then $f(A) = f(B)$.

Proof Suppose $A = B$, and let $y \in f(A)$. Then $\exists\, x \in A$, such that $y = f(x)$. Since $A = B$, then $x \in B$. Thus $y = f(x) \in f(B)$. Hence

$$f(A) \subseteq f(B). \tag{4.7}$$

Similarly, we have

$$f(B) \subseteq f(A). \tag{4.8}$$

From Equations (4.7) and (4.8), one has $f(A) = f(B)$.

Remark 4.4.5 Even though $f(A) = f(B)$, it is not necessary that $A = B$. See Example 4.4.6 below.

Example 4.4.6 Let $X = \mathbb{Z}$, $Y = \mathbb{R}$ and let

$$f : X \to Y$$

be a function defined by $f(x) = x^2$. Let $A = \{-2, 1, 3\}$, $B = \{2, 1, -3\}$, then

$$f(A) = \{4, 1, 9\} \text{ and } f(B) = \{4, 1, 9\}.$$

Notice that $f(A) = f(B)$, but $A \neq B$.

Theorem 4.4.7 Let $f : X \to Y$ be a function and f^* be a relation from $P(X)$ to $P(Y)$ defined by

$$f^* = \{(A, B) \in P(X) \times P(Y) : f(A) = B\}.$$

Then f^* is a function from $P(X)$ to $P(Y)$, where $P(Z)$ is the power set of a set Z.

Proof First, suppose $A \in P(X)$, then $A \subseteq X$. Thus, $f(A) \subseteq Y$ from which we get $f(A) \in P(Y)$. Suppose $f(A) = B$, then

$$\forall A \in P(X),\ \exists\, B \in P(Y), \text{ such that } (A, B) \in f^*.$$

Now, supposing $(A, B_1) \in f^* \wedge (A, B_2) \in f^*$, we obtain

$$B_1 = f^*(A) \wedge B_2 = f^*(A),$$

and hence, $B_1 = B_2$. Therefore, $f^* : P(X) \to P(Y)$ is a well-defined function.

One can have the following relations on direct images.

Theorem 4.4.8 If $f : A \to B$ is a mapping and C, D are subsets of A, then

(a) $f(C \cup D) = f(C) \cup f(D)$,

(b) $f(C \cap D) \subseteq f(C) \cap f(D)$,

(c) $f(C - D) \supseteq f(C) - f(D)$.

Proof We here prove parts (a) and (c), and leave the proof of part (b) to the readers as exercise.

(a) Suppose $y \in f(C \cup D)$. Then $y \in f(C \cup D) \Leftrightarrow \exists x \in C \cup D$, such that $y = f(x)$
$\Leftrightarrow \exists x \in C \wedge x \in D$, such that $y = f(x)$
$\Leftrightarrow (\exists x \in C, \text{ such that } y = f(x)) \wedge (\exists x \in D, \text{ such that } y = f(x))$
$\Leftrightarrow f(x) \in f(C) \wedge f(X) \in f(D)$
$\Leftrightarrow y \in f(C) \wedge y \in f(D)$
$\Leftrightarrow y \in (f(C) \cup f(D))$.
Consequently, $f(C \cup D) = f(C) \cup f(D)$.

(c) Suppose $y \in f(C) - f(D)$. Then $y \in f(C) \wedge y \notin f(D)$. Since $y \in f(C), \exists x \in C$, such that $y = f(x)$. This implies

$$y = f(x) \notin f(D), \text{ since } y \notin f(D).$$

Therefore, $x \notin D$; i.e.,
$$\exists x \in C \wedge x \notin D$$
such that
$$y = f(x).$$
So, $\exists x \in C - D$ such that $y = f(x)$. This implies that $y \in f(C - D)$. In conclusion,
$$y \in f(C) - f(D) \Rightarrow y \in f(C - D).$$
Thus, $f(C) - f(D) \subseteq f(C - D)$.

Remark 4.4.9 The following example shows that

$$f(C) \cap f(D) \nsubseteq f(C \cap D).$$

Let $A = \{3,5\}, B = \{2\}$, and let

$$f : A \to B$$

be constant function. Suppose $C = \{3\}, D = \{5\}$. Then

$$f(C \cap D) = f(\phi) = \phi,$$

and

$$f(C) = f(\{3\}) = \{2\}, \quad f(D) = f(\{5\}) = \{2\}.$$

Thus,

$$f(C) \cap f(D) = \{2\},$$

from which we obtain that

$$f(C) \cap f(D) \nsubseteq f(C \cap D).$$

Definition 4.4.10 Let $f : A \rightarrow B$ be a function, and let $D \subseteq B$. The set

$$f^{-1}(D) = \{x \in A : f(x) \in D\}$$

is called the inverse image (or, the pre image) of D under f.

Note that, even though the inverse images are expressed under the symbol f^{-1}, this symbol does not mean the inverse function of f. It is simply used pure symbolically. Even if f is not invertible, one can define and write the corresponding inverse images as above.

Example 4.4.11 Let $f : \mathbb{R} \rightarrow \mathbb{R}$ be a function defined by

$$f(x) = x^2 + 1,$$

then

$$
\begin{aligned}
f^{-1}(17) &= \{x \in \mathbb{R} : f(x) = 17\} \\
&= \{x \in \mathbb{R} : x^2 + 1 = 17\} \\
&= \{x \in R : x^2 = 16\} \\
&= \{-4, 4\}.
\end{aligned}
$$

Also

$$
\begin{aligned}
f^{-1}(\{5, 10\}) &= \{x \in A : f(x) \in \{5, 10\} \\
&= \{x \in A : x^2 + 1 = 5 \vee x^2 + 1 = 10\} \\
&= \{-2, 2, -3, 3\}.
\end{aligned}
$$

Theorem 4.4.12 Let $f : X \rightarrow Y$ be a function and C, D be subsets of Y. If $C = D$, then $f^{-1}(C) = f^{-1}(D)$.

Proof Suppose $C = D$ and let $x \in f^{-1}(C)$. Then

$$
\begin{aligned}
x \in f^{-1}(C) &\Leftrightarrow f(x) \in C \\
&\Leftrightarrow f(x) \in D (\text{since } C = D) \\
&\Leftrightarrow x \in f^{-1}(D).
\end{aligned}
$$

Hence, $f^{-1}(C) = f^{-1}(D)$.

Remark 4.4.13 Even though $f^{-1}(C) = f^{-1}(D)$, it is not necessary that $C = D$. See Example 4.4.14 below.

Example 4.4.14 Let $f : \mathbb{R} \rightarrow \mathbb{R}$ be a function defined by

$$f(x) = |x|,$$

and let $C = (0,1)$, $D = (-2,1)$. Then

$$f^{-1}((0,1)) = (-1,1),$$
$$f^{-1}((-2,1)) = (-1,1).$$

Notice that $f^{-1}((-2,1)) = f^{-1}((-1,1))$, but $(-2,1) \neq (-1,1)$.

Theorem 4.4.15 Let $f : A \to B$ be a function and C, D be subsets of B. Then

(a) $f^{-1}(C \cup D) = f^{-1}(C) \cup f^{-1}(D)$,

(b) $f^{-1}(C \cap D) = f^{-1}(C) \cap f^{-1}(D)$,

(c) $f^{-1}(C - D) = f^{-1}(C) - f^{-1}(D)$.

Proof We show part (b) and leave the proof of parts (a) and (c) to the readers as exercise.

(b) Suppose $x \in f^{-1}(C \cap D)$. Then

$$x \in f^{-1}(C \cap D) \Leftrightarrow f(x) \in C \cap D$$
$$\Leftrightarrow f(x) \in C \wedge f(x) \in (D)$$
$$\Leftrightarrow x \in f^{-1}(C) \wedge x \in f^{-1}(D)$$
$$\Leftrightarrow x \in f^{-1}(C) \cap f^{-1}(D).$$

Thus $f^{-1}(C \cap D) = f^{-1}(C) \cap f^{-1}(D)$.

Notion 4.4.16 For a set A, by 2^A, we denote the set of all the functions from A to $\{a, b\}$.

Theorem 4.4.17 For every set A, then there exists a bijection between 2^A and the power set $P(A)$ of A.

Proof Suppose we let $2 = \{0, 1\}$, notationally. We define a relation R from $P(A)$ to 2^A by

$$R = \{(B, F) \in P(A) \times 2^A : f = \lambda_B\},$$

where λ_B is the characteristic function, i.e.,

$$\lambda_B(x) = \begin{cases} 0, & x \in B, \\ 1, & x \in A - B, \end{cases}$$

First, we prove that R is a function from $P(A)$ to 2^A. Assuming that $B \in P(A)$, we obtain $\exists \, \lambda_B \in 2^A$. Thus,
$\forall \, B \in P(A)$, $\exists \, \lambda_B \in 2^A$, so that

$$(B, \lambda_B) \in R. \tag{4.9}$$

Let $(B, f_1) \in R$, $(B, f_2) \in R$. Then $f_1 = \lambda_B \wedge f_2 = \lambda_B$. Hence $f_1 = f_2$.
Therefore, by Equation (4.9),

$$R : P(A) \to 2^A$$

is a function.

Now, we show that R is one-to-one. To do that, we let $B_1, B_2 \in P(A)$, satisfying $R(B_1) = R(B_2)$. Since $R(B_1) = \lambda_{B_1}$, $R(B_2) = \lambda_{B_2}$, then $\lambda_{B_1} = \lambda_{B_2}$. Thus,

$$B_1 = \{x \in A : \lambda_{B_1}(x) = 0\} = \{x \in A : \lambda_{B_2}(x) = 0\} = B_2.$$

So, $R : P(A) \to 2^A$ is one-one function.

Finally, we prove that R is onto function. Suppose $f \in 2^A$, then $f : A \to 2 = \{0, 1\}$. Assuming $B = f^{-1}(0)$, we obtain that $B \in P(A)$, but $R(B) = \lambda_B$. Since

$$f(x) = \begin{cases} 0, & x \in B, \\ 1, & x \in A - B, \end{cases}$$

then $R(B) = \lambda_B = f$. So, $R : P(A) \to 2^A$ is onto.

As R is both one-to-one and onto, it is a bijection from $P(A)$ to 2^A.

Exercises

1. Let $X = \{a, b, c, d\}$, $Y = \{x, y, z, w\}$, and let
   $f : X \to Y$ be a function defined by
   $$f(a) = x, \qquad f(b) = y$$
   $$f(c) = z, \qquad f(d) = w.$$
   Find each of
   $f(\{a, b\}), f(\{b, c\}), f(\{c, d\})$
   $f^{-1}(\{x, z\}), f^{-1}(\{y, z, w\})$.

2. Let $f : \mathbb{R} \to \mathbb{R}$ be defined by $f(x) = 2x^2 + 1$. Find $f((0, 2))$.

3. Let $f : \mathbb{R} \to \mathbb{R}$ be defined by $f(x) = \frac{x}{2} - 1$. Find $f((-1, 1))$.

4. Let $f : \mathbb{R} \to \mathbb{R}$ be defined by

$$f(x) = \begin{cases} x - 2, & x \geq 2, \\ 2x + 1, & x < 2. \end{cases}$$

   Find $f^{-1}((1, 3))$.

5. Let $f : \mathbb{R} \to \mathbb{R}$ be defined by $f(x) = x^2 - 6x + 9$. Find and sketch each of
   $f([0, 1)), f([2, 4)), f((0, 3))$
   $f^{-1}([-2, 0]), f^{-1}((1, 3)), f^{-1}([0, 1))$.

6. Let $f : A \to B$ be a function and let $C \subseteq A$, $D \subseteq B$.

   Prove each of the following assertions:

   (a) $C \subseteq f^{-1}(f(C))$,
   (b) $f(f^{-1}(D)) \subseteq D$,
   (c) $C = f^{-1}(f(C))$ if f is one–one,
   (d) $f(f^{-1}(D)) = D$ if f is onto.

7. Let $f : A \to B$ be a function. Prove that if f is a bijection, then $f^* : P(A) \to P(B)$ is a bijection.

8. Let $f : \mathbb{R} \times \mathbb{R} \to \mathbb{R}$ be a function defined by

   $$f(x,y) = x^2 + y^2.$$

   (a) Find $f^{-1}([0,1])$.
   (b) Find $\text{Rng}(f)$.
   (c) Find $f(A)$, if $A = \{(x,y) : x+y = \sqrt{2}\}$.

9. Let X be a nonempty set. Define a relation R on X^X by

   $$f R g \text{ if and only if } \text{Im}(f) = \text{Im}(g).$$

   Prove that R is an equivalence relation on X^X.

10. Let $f : \mathbb{N} \times \mathbb{N} \to \mathbb{N}$ be a function defined by

    $$f(m,n) = 2^m(2n+1).$$

    Find

    (a) $f^{-1}(\{1,2,3,4,5,6\})$.
    (b) $f^{-1}(\{4,6,8,10\})$.
    (c) $f(\{(1,1),(2,2),(4,1),(1,4)\})$.

11. Let $f : \mathbb{N} \times \mathbb{N} \to \mathbb{N}$ be a function defined by $f(m,n) = 2^m 3^n$.

    Find

    (a) $f(A \times B)$, where $A = \{1,2,4\}$, $B = \{3,4\}$,
    (b) $f^{-1}(\{5,6,7,8,9,10\})$.

12. Let $f : \mathbb{N} \times \mathbb{N} \to N$ be a function defined by

    $$f(m,n) = 2^{m-1}(2n-1).$$

    Show that f is one-one and onto N.

5 Cardinality of Sets

Let A and B be sets. It is natural to ask whether or not A and B have the same number of elements. In this chapter, we concentrate on studying when the given two sets A and B have the same number of elements.

If both A and B are finite sets, it is relatively easy to check these sets have the same number of elements, by directly counting the number of elements of the sets. But this counting technique cannot be used for infinite sets. In such a case, we then try to find a function,

$$f : A \rightarrow B,$$

which is both one-to-one and onto. That is, if there exists at least one one-to-one correspondence, or a bijection f from A onto B, then one can verify that the two sets A and B have the same number of elements.

5.1 CARDINAL NUMBER

5.1.1 EQUIPOTENT SETS

We here concentrate on considering given two sets have the same number of elements by the existence of at least one bijection between them.

Definition 5.1.1 Two sets A and B are said to be equipotent, denoted by $A \approx B$, if there exists at least one bijection (or one one-to-one correspondence),

$$f : A \rightarrow B, \text{ or } f : B \rightarrow A,$$

which is both one-to-one and onto. If there are no bijections between two sets, then A and B are said to be not equipotent, and we write $A \not\approx B$.

As we considered in Chapter 4, if a function $f : A \rightarrow B$ is a bijection, then its inverse function naturally exists,

$$f^{-1} : B \rightarrow A.$$

So, the above definition can be simply re-defined by that: $A \approx B$, if there exists a bijection,

$$f : A \rightarrow B.$$

Example 5.1.2 Let $A = \{a,b,c,d\}$ and $B = \{1,4,7,10\}$, and let

$$f : A \rightarrow B$$

be the function defined by

$$f(a) = 1, f(b) = 4, f(c) = 7, f(d) = 10.$$

It is clear that f is a bijection, hence $A \approx B$.

DOI: 10.1201/9780429022838-5

Example 5.1.3 Let $X = \{x,y,z\}$ and $Y = \{a,b\}$. It is impossible to find a bijection between X and Y. Hence X and Y are not equivalent; that is, $X \not\approx Y$.

Example 5.1.4 Let A and B be the open intervals (0, 1) and (2, 5), respectively. Let $f : A \to B$ be a function defined by

$$f(x) = 3x + 2.$$

Since f is a bijection, $A \approx B$.

Example 5.1.5 Let $\mathbb{N}$ be the set of all natural numbers, and $\mathbb{M}$ be the set of all even numbers, and let $f : \mathbb{N} \to \mathbb{M}$ be a function defined by

$$f(n) = 2n.$$

Note that f is a bijection. Consequently, $\mathbb{N} \approx \mathbb{M}$.

In the very above example, we note that $\mathbb{M} \subset \mathbb{N}$, and $\mathbb{N} \approx \mathbb{M}$, showing that set $\mathbb{N}$ of natural numbers is equipotent to a proper subset $\mathbb{M}$ of all even numbers. It provides a difference between finiteness and infiniteness of sets.

Definition 5.1.6 A set A is an infinite set, if it is equipotent to a proper subset of itself.

By Example 5.1.5 and Definition 5.1.6, one can conclude that, indeed, set $\mathbb{N}$ of natural numbers is infinite.

Theorem 5.1.7 The relation $\approx$ is an equivalence relation.

Proof Let A be a set. Then there exists a well-defined function,

$$id_A : A \to A,$$

defined by

$$id_A(a) = a, \text{ for all } a \in A.$$

It is not hard to check this function id_A is a bijection.

Suppose now two sets A and B are equipotent. Since $A \approx B$, there exists a bijection,

$$f : A \to B,$$

implying the existence of a bijection,

$$f^{-1} : B \to A,$$

where f^{-1} is the inverse of f. It shows that $B \approx A$.

Now, assume that sets A and B are equipotent, and the sets B and C are equipotent, that is,

$$A \approx B, \text{ and } B \approx C.$$

Then there exists bijections,

$$g : A \to B, \text{ and } h : B \to C.$$

So, one can define the composition,

$$h \circ g : A \to C,$$

of h and g. Note and recall that this function $h \circ g$ is not only invertible, but also,

$$(h \circ g)^{-1} = g^{-1} \circ h^{-1} : C \to A,$$

implying the bijectivity of $h \circ g$. Thus,

$$A \approx C.$$

Therefore, the relation $\approx$ is an equivalence relation.

The equipotent property, or the equipotence, also satisfies the following result as an equivalence relation.

Theorem 5.1.8 Let A, B, C, and D be sets with $A \approx C$ and $B \approx D$. If A and B are disjoint, and if C and D are disjoint, then $A \cup B \approx C \cup D$.

Proof Since $A \approx C$, then there is a bijection $f : A \to C$, and since $B \approx D$, then there is a bijection $g : B \to D$. Then one can define a bijection,

$$f \cup g : A \cup B \to C \cup D,$$

defined by

$$f \cup g (x) = \begin{cases} f(x) & \text{if } x \in A \\ g(x) & \text{if } x \in B, \end{cases}$$

for all $x \in A \cup B$. Note that, by definition, the function,

$$f \cup g,$$

is surjective; and by the disjoint conditions, it is injective, and hence, it is a bijection. Therefore, $A \cup B \approx C \cup D$.

Similarly, one can get the following result.

Theorem 5.1.9 Let A, B, C, and D be sets with $A \approx C$ and $B \approx D$. Then $A \times B \approx C \times D$.

Proof One can define a bijection,

$$f \times g : A \times B \to C \times D,$$

by

$$f \times g ((a, b)) = (f(a), g(b)),$$

for all $(a, b) \in A \times B$. We leave the detailed proof to the readers.

Definition 5.1.10 The number of elements of a set A is called the cardinality of A. The cardinality of A is denoted by $\#(A)$.

Notion 5.1.11 The cardinal number of ϕ is 0, denoted by $\#(\phi) = 0$.
   The cardinality of $\{1\}$ is 1, denoted by $\#(\{1\}) = 1$.
   The cardinality of $\{1,2\}$ is 2, denoted by $\#(\{1,2\}) = 2$.
   $\vdots$
   The cardinality of $\{1,2,3,\ldots,k\}$ is k, denoted by

$$\#(\{1,2,3,\ldots,k\}) = k.$$

From below, we use symbol $\mathbb{N}_k$ to write the set $\{1,2,3,\ldots,k\}$.

Theorem 5.1.12 The fundamental property of the cardinalities is that:

$$A \approx B \iff \#(A) = \#(B).$$

Proof Indeed, if $A \approx B$, then there exists a bijection $f : A \to B$, which is both one-to-one and onto. By the one-to-one-ness and the onto-ness of f,

$$\#(A) = \#(B).$$

Conversely, if $\#(A) = \#(B)$, then one can construct a bijection between A and B because of the one-to-one-ness and the onto-ness of bijections, and hence, $A \approx B$. We will consider it more precisely below.
   By this equivalence, one may/can understand the cardinality, say α, which is a quantity, as a family of all sets having the same cardinalities α. For instance,

$$\{a,b\},\ \{1,2\},\ \{0,-2\},\ \{x,y\} \in 2,\ \text{etc.},$$

by regarding 2 as a family of all sets having their cardinalities 2. That is, the quantity 2 represents the cardinalities of all sets having two elements.

Definition 5.1.13 A cardinal number α is the quantity, representing the cardinalities of all equipotent sets having α-many elements.

Remark 5.1.14

1. The cardinal numbers induced by finite sets are nothing but the cardinalities of the sets.

2. The set A is finite if there does not exist a proper subset B of A, such that $B \approx A$. So, the finite cardinal numbers are determined uniquely up to cardinalities.

3. The cardinal number α is a finite number if it is the cardinality of a finite set. Otherwise it is called an infinite cardinal number. In such a sense, a finite cardinal number is also called a natural number. An infinite cardinal number is called a transfinite number.

4. A set S is finite, if and only if

$$\#(S) \neq \#(S) + 1,$$

so, if $\alpha = \#(S)$ and S is a finite set, then

$$\alpha \neq \alpha + 1.$$

Theorem 5.1.15 Let A and B be finite sets. Then

$$\#(A \cup B) = \#(A) + \#(B) - \#(A \cap B).$$

Meanwhile, the equality

$$\#(A \cup B) = \#(A) + \#(B)$$

holds true, if and only if A and B are disjoint in the sense that

$$A \cap B = \phi.$$

Proof Suppose two sets A and B are finite sets, and hence,

$$\#(A), \ \#(B) \in \mathbb{N}.$$

We let

$$n = \#(A), \text{ and } k = \#(B),$$

for $n, k \in \mathbb{N}$. Then the union $X = A \cup B$,

$$X = \{x \mid x \in A, \text{ or } x \in B\},$$

and the intersection $Y = A \cap B$,

$$Y = \{y \mid y \in A, \text{ and } y \in B\},$$

are well-defined as new finite sets.

Observe that

$$\#(X) \leq n + k,$$

since the quantity $n + k$ double-counts the cardinality $\#(Y)$. Therefore,

$$\#(X) + \#(Y) = n + k,$$

if and only if

$$\#(X) = n + k - \#(Y).$$

Meanwhile, if the intersection Y is empty, that is, $Y = \phi$, then the above general formula satisfies that

$$\#(X) = n + k - 0 = n + k,$$

since $\#(\phi) = 0$. Conversely, if Y is not empty, that is, $Y \neq \phi$, then

$$\#(X) = n + k - \#(Y) \neq n + k,$$

because $\#(Y) \neq 0 = \#(\phi)$.

Now, let's consider the cases where cardinal numbers are not finite.

Definition 5.1.16 The cardinal number $\aleph$ is defined to be the cardinality $\#(\mathbb{N})$ of N, that is,

$$\aleph = \#(\mathbb{N}).$$

Sometimes, $\aleph$ is called the countable infinity.

Definition 5.1.17 The cardinal number c is defined to be the cardinality $\#((0,1))$ of the subset $(0,1)$, the open interval,

$$(0,1) = \{r \in \mathbb{R} \mid 0 < r < 1\},$$

of the set $\mathbb{R}$ of all real numbers. This cardinal number c is called the continuum (or, the uncountable infinity).

Example 5.1.18 Let (a,b) be any open interval in $\mathbb{R}$, and let

$$f : (0,1) \rightarrow (a,b)$$

be a function defined by $f(x) = a + (b - a)x$. Then this function f is a bijection. Thus, (a,b) has cardinality c.

Example 5.1.19 The function $f : (-\frac{\pi}{2}, \frac{\pi}{2}) \rightarrow \mathbb{R}$ defined by

$$f(x) = \tan x,$$

is a bijection. Hence the set $\mathbb{R}$ of real numbers has its cardinality, the continuum c, that is,

$$\#(\mathbb{R}) = c.$$

So, without loss of generality, one can re-define the cardinal number c, the continuum, by the cardinality of $\mathbb{R}$.

Theorem 5.1.20 The transfinite number $\aleph$ is strictly less than c. That is,

$$\aleph < c,$$

as cardinal numbers. And there are no cardinal numbers between $\aleph$ and c.

Proof One can find a one-to-one function (or, an injection),

$$f : \mathbb{N} \rightarrow \mathbb{R},$$

by

$$f(n) = n \text{ in } \mathbb{R}, \text{ for all } n \in \mathbb{N}.$$

i.e., this embedding function f is clearly one-to-one. It means that the range of f is a proper subset of $\mathbb{R}$, and hence,

$$\#(\mathbb{N}) = \aleph \leq c = \#(\mathbb{R}).$$

Moreover, there are no one-to-one functions from $\mathbb{R}$ to $\mathbb{N}$ (See Section 5.1.2 further: one can find a bijection from subset $\mathbb{Q}$ of all rational numbers to $\mathbb{N}$). It shows that

$$c \neq \aleph.$$

Therefore,

$$\aleph < c,$$

as transfinite cardinal numbers.

Logically, there is no cardinal number between $\aleph$ and c. Indeed, if there were a cardinal number β such that

$$\aleph \leq \beta < c,$$

then

$$\beta = \aleph + k, \text{ for some } k \in \{0\} \cup \mathbb{N} \cup \{\aleph\},$$

implying that

$$\beta = \#(A) = \#(\mathbb{N}) = \aleph,$$

by the equipotent property of A and $\mathbb{N}$:

$$A \approx \mathbb{N} \cup \mathbb{N}_k \approx \mathbb{N}.$$

Also, if there were a cardinal number γ satisfying

$$\aleph < \gamma \leq c,$$

then

$$\gamma = c - k, \text{ for some } l \in \{0\} \cup \mathbb{N} \cup \{\aleph\},$$

implying that

$$\gamma = \#(B) = \#(\mathbb{R}) = c,$$

by the equipotent property of B and $\mathbb{R}$:

$$B \approx \mathbb{R} - \mathbb{N}_k \approx \mathbb{R}.$$

Therefore, there are no cardinal numbers between $\aleph$ and c.

The above theorem shows that there are only two transfinite (or infinite) cardinal numbers $\aleph$ and c. By discussions, if we define set $\mathscr{C}$ of all cardinal numbers, then

$$\mathscr{C} = \{0\} \cup \mathbb{N} \cup \{\aleph\} \cup \{c\}.$$

In particular, all finite cardinal numbers are from subset $\{0\} \cup \mathbb{N}$ of $\mathscr{C}$, and the transfinite cardinal numbers are $\{\aleph, c\}$ in $\mathscr{C}$.

5.1.2 COUNTABLE SETS

In the previous section, we consider the equipotent property on sets, and the cardinality of sets, and show that two sets are equipotent, if and only if they have the same cardinality. Based on the cardinalities of sets, we define and study cardinal numbers,

$$\{0\} \cup \mathbb{N} \cup \{\aleph\} \cup \{c\}.$$

In this section, we study a transfinite cardinal number, $\aleph$, more in detail. Recall that

$$\aleph = \#(\mathbb{N}).$$

Definition 5.1.21 A set A is said to be countable (or, denumerable), if

$$\#(A) \leq \aleph.$$

In particular, if a set A is a finite set, then it is said to be finitely countable, meanwhile, if it is equipotent to set $\mathbb{N}$ of natural numbers, then A is said to be infinitely countable.

By definition, the empty set ϕ whose cardinality is 0, and all nonempty finite sets whose cardinalities are in $\mathbb{N}$, and the sets equipotent to $\mathbb{N}$ whose cardinalities are $\aleph$ are countable.

Definition 5.1.22 A set is said to be uncountable (or, not denumerable), if it is not countable.

By definition, every set equipotent to $\mathbb{R}$, equivalently, whose cardinality is the continuum c, is uncountable.

Also, by the above definitions, one can realize that there are two kinds of infinity, denoted usually by ∞; the countable infinity $\aleph$, and the uncountable infinity c, satisfying

$$\aleph < c,$$

in terms of transfinite numbers.

Theorem 5.1.23 Set $\mathbb{Z}$ of all integers is countable.

Proof There exists a bijection $f : \mathbb{N} \to \mathbb{Z}$ defined by

$$f(x) = \begin{cases} \frac{x}{2}, & \text{if } x \text{ is even,} \\ \frac{1-x}{2}, & \text{if } x \text{ is odd.} \end{cases}$$

We leave the detailed proof to the readers.

Theorem 5.1.24 The Cartesian product $\mathbb{N} \times \mathbb{N}$ is countable.

Proof There is a bijection $f : \mathbb{N} \times \mathbb{N} \to \mathbb{N}$ by

$$f(m,n) = 2^{m-1}(2n-1).$$

We leave the detailed proof to the readers.

Example 5.1.25 Any infinite sequence,

$$(a_1, a_2, a_3, \ldots)$$

of mutually distinct elements is countable (as a set), since there is a bijection,

$$f(n) = a_n,$$

whose domain is $\mathbb{N}$. Note here that the mutual distinctness of entries is crucial.

Theorem 5.1.26 Set $\mathbb{Q}^+ = \{q \in \mathbb{Q} \mid q > 0\}$ of positive rational numbers is countable.

Proof Define a bijection $f : \mathbb{Q}^+ \to \mathbb{N} \times \mathbb{N}$ by

$$f(\frac{p}{q}) = (p, q), \text{ for all } \frac{p}{q} \in \mathbb{Q}^+,$$

where p and q are positive integers that are relatively prime in the sense that: $\gcd(p, q) = 1$, where gcd means the greatest common divisor. That is,

$$\mathbb{Q}^+ \approx \mathbb{N} \times \mathbb{N}.$$

We leave the detailed proof for the readers as an exercise.

Since $\mathbb{N} \times \mathbb{N} \approx \mathbb{N}$ by the very above theorem, we have

$$\mathbb{Q}^+ \approx \mathbb{N} \times \mathbb{N} \approx \mathbb{N}.$$

Therefore, $\mathbb{Q}^+$ is countable.

Theorem 5.1.27 If a set A is countable, then $A \cup \{x\}$ is countable, where x is an arbitrary element.

Proof Suppose $f : \mathbb{N} \to A$ is a bijection, making $\mathbb{N}$ and A are equipotent. If $x \in A$, then $A \cup \{x\} = A$, which is countable. If $x \notin A$, one can define a bijection $g : \mathbb{N} \to A \cup \{x\}$ by

$$g(n) = \begin{cases} x, & \text{if } n = 1, \\ f(n-1), & \text{if } n \neq 1. \end{cases}$$

We leave the detailed proof for the readers. The existence of this bijection g guarantees that

$$\mathbb{N} \approx A \cup \{x\},$$

implying the countability of $A \cup \{x\}$.

By above results, one obtains the following result.

Theorem 5.1.28 The set $\mathbb{Q}$ of all rational numbers is countable.

Proof It is shown that subset $\mathbb{Q}^+$ of $\mathbb{Q}$ is countable. Now, define a subset $\mathbb{Q}^{-1}$ of $\mathbb{Q}$ by

$$\mathbb{Q}^- = \{y \in \mathbb{Q} \mid y < 0\}.$$

Then there exists a bijection $h : \mathbb{Q}^+ \to \mathbb{Q}^-$, defined by

$$h(x) = -x, \text{ for all } x \in \mathbb{Q}^+.$$

It shows that subset $\mathbb{Q}^-$ is countable.

Also, by the very above theorem, a subset

$$\mathbb{Q}_0^+ = \mathbb{Q}^+ \cup \{0\}$$

of $\mathbb{Q}$ is countable.

Note-and-recall that $\mathbb{N} \approx E$, where E is the proper subset of $\mathbb{N}$ consisting of all even numbers, by the existence of a bijection $g_e : \mathbb{N} \to E$ defined by

$$g_e(n) = 2n, \text{ for all } n \in \mathbb{N}.$$

And, similarly, one can check that $\mathbb{N} \approx O$, where O is the proper subset of $\mathbb{N}$ consisting of all odd numbers, by the existence of a bijection $g_o : \mathbb{N} \to O$ defined by

$$g_o(n) = 2n - 1, \text{ for all } n \in \mathbb{N}.$$

Thus, one can have the equipotent properties,

$$\mathbb{Q}_0^+ \approx E \approx \mathbb{N}, \text{ and } \mathbb{Q}^- \approx O \approx \mathbb{N}.$$

So, one obtains that

$$\mathbb{Q} = \mathbb{Q}_0^+ \cup \mathbb{Q}^- \approx E \cup O = \mathbb{N}.$$

Therefore, set $\mathbb{Q}$ is countable.

The above theorem shows that

$$\#(\mathbb{Q}) = \aleph.$$

Theorem 5.1.29 Every subset of a countable set is countable.

Proof Suppose a set A is countable. Assume first that A is finite. If $A = \phi$, then it has only one subset ϕ, itself, satisfying $\#(\phi) = 0$. If $A \approx \mathbb{N}_k$, for some $k \in \mathbb{N}$, then

$$\#(B) \leq k < \aleph,$$

for all subsets B of A. So, if A is finitely countable, then all subsets of A are (finitely) countable.

Now, assume that A is infinitely countable, i.e., $A \approx \mathbb{N}$, equivalently, $\#(A) = \aleph$. Note that a subset X of $\mathbb{N}$ satisfies

$$X = \phi,$$

or
$$X \approx \mathbb{N}_k, \text{ for some } k \in \mathbb{N},$$

or
$$X = \mathbb{N},$$

implying that
$$\#(X) \in \{0\} \cup \mathbb{N} \cup \{\aleph\}.$$

Since $A \approx \mathbb{N}$, if B is an arbitrary subset of A, then
$$\#(B) \in \{0\} \cup \mathbb{N} \cup \{\aleph\},$$

too. Therefore, all subsets of a countable set A are countable.

Exercises

1. Prove that set $\mathbb{Z}$ of all integers is countable.

2. Let A and B be sets. Prove that if A is finite, then $A \cap B$ is finite.

3. Prove that $\mathbb{N}_k \times \mathbb{N}_r$ is finite, for $k, r \in \mathbb{N}$.

4. Prove that if A is infinite, and $A \subseteq B$, then B is infinite.

5. Prove that if A and B are finite sets, then $A \times B$ is finite.

6. Show that if $A \approx \phi$, then $A = \phi$.

7. Show that a set $S = \left\{ \frac{1}{2^k}, k \in \mathbb{N} \right\}$ is countable.

8. Prove that if A is countable, and B is finite, then set $A \cup B$ is countable.

9. Let A and B be countable sets. Prove that
$$A \times B \approx B \times A.$$

10. Verify that
    (a) $(0,1) \approx [0,1]$,
    (b) $(0,1) \approx (1,\infty)$,
    (c) $[0,1] \approx (0,1)$,
    (d) $[0,1] \approx [0,1)$.

11. Let A, B, and C be sets. Prove that
$$(A \times B) \times C \approx A \times (B \times C).$$

12. Let A and B be countable sets, such that
$$\#(B - C) = \#(C - B).$$
Prove that $\#(B) = \#(C)$.

13. Let A be countable infinite set, and $B \subseteq A$, such that $(A - B)$ be finite. Prove that $\#(A) = \#(B)$.

5.1.3   CARDINAL ARITHMETIC

We have showed that if A and B are finite sets, then

$$\#(A \cup B) = \#(A) + \#(B) - \#(A \cap B),$$

in $\mathbb{N} \cup \{0\}$. In this section, we consider such arithmetical operations on the set

$$\mathscr{C} = \{0\} \cup \mathbb{N} \cup \{\aleph\} \cup \{c\}$$

of all cardinal numbers.

First, the addition $(+)$ on $\mathscr{C}$ is formally defined as follows.

Definition 5.1.30 Let α and β be cardinals numbers, and let A and B be disjoint sets such that $\alpha = \#(A)$, $\beta = \#(B)$. Then

$$\alpha + \beta = \#(A \cup B).$$

Remark 5.1.31 The addition $(+)$ on $\mathscr{C}$ is well-defined. Indeed, let $A^* \approx A$ and $B^* \approx B$, such that

$$A \cap B = \phi \text{ and } A^* \cap B^* = \phi.$$

Then $A^* \cup B^* \approx A \cup B$. Hence

$$\#(A^* \cup B^*) = \#(A \cup B).$$

Therefore, the sum $\alpha + \beta$ is well-defined, where $\alpha = \#(A)$, $\beta = \#(B)$.

Example 5.1.32 Let $A = \{2,4,6\}$, $B = \{1,3,5,7,9\}$. Then

$$\alpha = \#(A) = 3, \text{ and } \beta = \#(B) = 5,$$

and hence,

$$3 + 5 = \alpha + \beta = \#\{1,2,3,4,5,6,7,9\} = 8,$$

since A and B are disjoint.

Example 5.1.33 Let

$$A = \{1,3,5,\dots\},$$
$$B = \{2,4,6,\dots\}.$$

Note that $\#(A) = \#(B) = \aleph$ and $A \cap B = \phi$. Then

$$\aleph + \aleph = \#(A \cup B)$$
$$= \#(\mathbb{N})$$
$$= \aleph,$$

implying that

$$\aleph + \aleph = \aleph.$$

Example 5.1.34 Let

$$S = (0,1) \subset R,$$
$$T = [1,2) \subset R.$$

Note that $\#(S) = \#(T) = c$ and $S \cap T = \phi$. Then

$$c + c = \#(S \cup T)$$
$$= \#((0,2))$$
$$= c.$$

Therefore,

$$c + c = c.$$

Theorem 5.1.35 Let α, β, and γ be cardinal numbers. Then

1. $(\alpha + \beta) + \gamma = \alpha + (\beta + \gamma)$,
2. $\alpha + \beta = \beta + \alpha$.

Proof

1. Let A, B, and C be mutually disjoint sets, such that

$$\alpha = \#(A), \beta = \#(B) \text{ and } \gamma = \#(C),$$

Then

$$(\alpha + \beta) + \gamma = \#(A \cup B) + \#(C)$$
$$= \#((A \cup B) \cup C)$$
$$= \#((A \cup (B \cup C))$$
$$= \#(A) + \#(B \cup C)$$
$$= \alpha + (\beta + \gamma).$$

2. If A and B are disjoint sets satisfying

$$\alpha = \#(A) \text{ and } \beta = \#(B),$$

then we have

$$\alpha + \beta = \#(A \cup B)$$
$$= \#(B \cup A)$$
$$= \beta + \alpha.$$

Remark 5.1.36 The cancellation law does not hold on $\mathscr{C}$. For example

$$\aleph + \aleph = \aleph = 1 + \aleph,$$

but $\aleph \neq 1$ in $\mathscr{C}$.

Now, let's consider the multiplication ($\cdot$) on $\mathscr{C}$.

Definition 5.1.37 Let α and β be cardinal numbers, and let A and B be such that $\alpha = \#(A)$, and $\beta = \#(B)$. Then the product $\alpha\beta$ of α and β is defined by

$$\alpha\beta = \#(A \times B).$$

Remark 5.1.38 The multiplication is well-defined. Indeed, suppose that $A^* \approx A$ and $B^* \approx B$. Then $A^* \times B^* \approx A \times B$. So,

$$\#(A^* \times B^*) = \#(A \times B).$$

This implies that $\alpha\beta$ is well-defined.

Note that, different from the addition ($+$), the multiplication ($\cdot$) does not need the disjoint condition by the very definition of the Cartesian product.

Example 5.1.39 Let $A = \{2,3,4\}, B = \{q,p,r,s,t\}$. Then

$$\alpha = \#(A) = 3, \beta = \#(B) = 5,$$

and

$$(3)(5) = \alpha\beta = \#(\{2,3,4\} \times \{q,p,r,s,t\}) = 15.$$

Theorem 5.1.40 Let α, β, and γ be cardinal numbers. Then

1. $\alpha(\beta\gamma) = (\alpha\beta)\gamma$,

2. $\alpha\beta = \beta\alpha$,

3. $\alpha(\beta + \gamma) = \alpha\beta + \alpha\gamma$.

Proof Let A, B, and C be sets satisfying

$$\alpha = \#(A), \beta = \#(B), \text{ and } \gamma = \#(C).$$

Then
$$\begin{aligned}
\alpha(\beta\gamma) &= \#(A \times (B \times C)) \\
&= \#(A \times B \times C) \\
&= \#((A \times B) \times C) = (\alpha\beta)\gamma.
\end{aligned}$$
Thus, the first formula holds.
    Since
$$A \times B \approx B \times A,$$
we have
$$\#(A \times B) = \alpha\beta = \beta\alpha = \#(B \times A),$$
and hence, the second formula holds.

Let A, B, and C be sets, satisfying

$$\alpha = \#(A), \beta = \#(B), \text{ and } \gamma = \#(C),$$

and assume further that the two sets B and C are disjoint. Then

$$\begin{aligned} \alpha(\beta + \gamma) &= \#(A)\#(B \cup C) \\ &= \#(A \times (B \cup C)) \\ &= \#[(A \times B) \cup (A \times C)]. \end{aligned}$$

Since $B \cap C = \phi$, we have

$$(A \times B) \cap (A \times C) = \phi.$$

Thus,

$$\begin{aligned} \alpha(\beta + \gamma) &= \#(A \times B) + \#(A \times C) \\ &= \alpha\beta + \alpha\gamma, \end{aligned}$$

implying the third formula.

Recall that a set of all functions from set B to set A is denoted by A^B.

Definition 5.1.41 Let α and β be cardinal numbers, and let A and B be sets satisfying that $\alpha = \#(A)$, and $\beta = \#(B)$. Then the power α^β is defined by

$$\alpha^\beta = \# \left(A^B \right).$$

Remark 5.1.42 The power is well-defined on $\mathscr{C}$. Indeed, if $A_1 \approx A_2$ and $B_1 \approx B_2$, then $A_1^{B_1} \approx A_2^{B_2}$. Hence

$$\# \left(A_1^{B_1} \right) = \#(A_2^{B_2}),$$

implying that α^β is well-defined in $\mathscr{C}$.

Theorem 5.1.43 Let α, β, and γ be cardinal numbers of $\mathscr{C}$. Then

1. $\alpha^\beta \alpha^\gamma = \alpha^{\beta + \gamma}$,

2. $(\alpha^\beta)^\gamma = \alpha^{\beta\gamma}$,

3. $\alpha^\gamma \beta^\gamma = (\alpha\beta)^\gamma$.

Proof Let A, B, and C be sets, such that

$$B \cap C = \phi,$$

and

$$\alpha = \#(A), \beta = \#(B), \gamma = \#(C).$$

Note that

$$\alpha^{\beta+\gamma} = \#(A^{B\cup C}), \quad \alpha^{\beta}\alpha^{\gamma} = \#(A^B \times A^C),$$

and $A^{B\cup C}$ consists of all the functions with domain $B \cup C$ and codomain A. Also A^B and A^C have the similar meaning. So, it suffices to show

$$A^{B\cup C} \approx A^B \times A^C.$$

Let $f \in A^{B\cup C}$ corresponds to the ordered pair of functions,

$$(f|_B, f|_C).$$

Note that $(f|_B, f|_C)$ belongs to $A^B \times A^C$. The function

$$F : A^{B\cup C} \to A^B \times A^C$$

defined by

$$F(f) = (f|_B, f|_C),$$

is a bijection, because $B \cap C = \phi$. Thus,

$$A^{B\cup C} \approx A^B \times A^C.$$

Therefore, the first formula holds.

We leave the proof of the second and third formulas as exercise for the readers.

Recall that if A is a set, the corresponding power set $P(A)$ is defined to be the set of all subsets of A.

Theorem 5.1.44 If $\alpha = \#(A) \in \mathscr{C}$, then

$$\#(P(A)) = 2^{\alpha}.$$

Proof We already proved that there exists a bijection between $P(A)$ and 2^A. That is, $P(A) \approx 2^A$. Therefore,

$$\#(P(A)) = \#(2^A) = 2^{\alpha}.$$

Remark 5.1.45   1. The above theorem can be re-stated by that: $\#(P(A)) = 2^{\#(A)}$, for all sets A.

     2. Cantor's theorem tell us that, for any cardinal number α,

$$\alpha < 2^{\alpha}.$$

     3. $\aleph < 2^{\aleph}$.

     4. $2^{\aleph} = c$.

Theorem 5.1.46 On the set $\mathscr{C}$ of cardinal numbers, we have

1. $\aleph \aleph = \aleph$,

2. $\aleph c = c$,

3. $cc = c$.

Proof

1. Since $\mathbb{N} \times \mathbb{N}$ is countable,
$$\mathbb{N} \times \mathbb{N} \approx \mathbb{N}.$$
   Therefore, $\aleph \aleph = \aleph$.

2. Let $f : \mathbb{N} \times (0,1) \to (0,\infty)$ be a function defined by
$$f(x,y) = x+y.$$
   It is not difficult to show that f is a bijection. So,
$$\mathbb{N} \times (0,1) \approx (0,\infty).$$
   Hence, $\#(\mathbb{N} \times (0,1)) = \#(0,\infty)$. Therefore, $\aleph c = c$.

3. We have $cc = \#(\{(x,y) : x,y \in (0,1)\})$ and x,y can be written in the digit form of infinite decimals as follows:
$$x = 0.x_1 x_2 x_3 \ldots,$$
$$y = 0.y_1 y_2 y_3 \ldots.$$
   Note that $z = 0.x_1 y_1 x_2 y_2 \ldots$ is an element in $(0,1)$, too. So we have defined a one-to-one function,
$$f : (0,1) \times (0,1) \to (0,1).$$
   Conversely, one can define a one-to-one function,
$$g : (0,1) \to (0,1) \times (0,1).$$
   So,
$$(0,1) \times (0,1) \approx (0,1).$$
   i.e., we have
$$\#((0,1) \times (0,1)) = \#(0,1).$$
   Therefore, $cc = c$.

5.1.4   ORDER TYPES

The set $\mathscr{C}$ of all cardinal numbers is ordered in the following manner.

Definition 5.1.47 Let α and β be cardinal numbers. We say that $\alpha \leq \beta$, (which is read α is less than or equal to β), if there exists a one-to-one function $f : A \to B$, where $\alpha = \#(A)$ and $\beta = \#(B)$.

Example 5.1.48 Since $\mathbb{N}$ is a proper subset of $\mathbb{R}$,

$$\#(\mathbb{N}) = \aleph \leq c = \#(\mathbb{R}).$$

Furthermore, since $\mathbb{N}$ is countable, and $\mathbb{R}$ is uncountable,

$$\aleph < c.$$

Example 5.1.49 Let A, B, and C be sets, such that

$$\alpha = \#(A), \beta = \#(B), \gamma = \#(C).$$

If $f : A \to B$ is one-to-one, and $g : B \to C$ is one-to-one, then $g \circ f : A \to C$ is one-to-one. Therefore

$$\text{if } \alpha \leq \beta, \text{ and } \beta \leq \gamma, \text{ then } \alpha \leq \gamma.$$

Example 5.1.50 For every cardinal number $\alpha \in \mathscr{C}$,

$$\alpha \leq \alpha,$$

since every set is a subset of itself.

In view of the preceding examples, the following theorem holds true.

Theorem 5.1.51 For cardinal numbers α, β, and γ, the following properties hold:

1. $\alpha \leq \alpha$ (Reflexiveness).

2. If $\alpha \leq \beta$ and $\beta \leq \alpha$, then $\alpha = \beta$ (Anti-symmetry).

3. If $\alpha \leq \beta$ and $\beta \leq \gamma$, then $\alpha \leq \gamma$ (Transitivity).

Proof By definition, for all $\alpha \in \mathscr{C}$,

$$\alpha \leq \alpha \text{ in } \mathscr{C}.$$

Suppose $\alpha, \beta \in \mathscr{C}$, and assume that

$$\alpha \leq \beta, \text{ and } \beta \leq \alpha.$$

If $\alpha = \#(A)$, and $\beta = \#(B)$, for sets A and B, then the above order relations (or, inequalities) mean that

$$f(A) \subseteq B, \text{ and } f^{-1}(B) \subseteq A,$$

where $f : A \to B$ is an one-to-one function, and

$$f^{-1}(B) = \{a \in A \mid f(a) \in B\}$$

is the pre image of f. By the one-to-one-ness of f and the second set inclusion, one can realize that the pre image $f^{-1}(B)$ is identified with the range of the inverse $f^{-1} : B \to A$, implying that this one-to-one function f is actually a bijection. In other words,

$$A \approx B, \text{ if and only if } \#(A) = \#(B).$$

Therefore, two cardinal numbers α and β are identically same in $\mathscr{C}$.

Now, assume that

$$\alpha \leq \beta \text{ and } \beta \leq \gamma,$$

in $\mathscr{C}$. If $\alpha = \#(A)$, $\beta = \#(B)$, and $\gamma = \#(C)$ for sets A, B, and C, respectively, then the above inequalities implies that

$$f(A) \subseteq B \text{ and } g(B) \subseteq C,$$

for one-to-one functions $f : A \to B$ and $g : B \to C$. So, there exists a one-to-one function,

$$g \circ f : A \to C,$$

implying that

$$g \circ f(A) \subseteq C.$$

By the one-to-one-ness of $g \circ f$, we have

$$\#(A) \leq \#(C), \text{ if and only if } \alpha \leq \gamma.$$

5.1.5   MORE ABOUT CARDINAL NUMBERS

In this section, we study more relations between cardinal numbers and cardinalities.

Theorem 5.1.52 (Cantor's theorem) If A is a set, then

$$\#(A) < \#(P(A)).$$

Proof Let $g : A \to P(A)$ be a function defined by

$$g(a) = \{a\}.$$

It is clear that g is one-to-one. Hence

$$\#(A) \leq \#(P(A)).$$

Thus, it is sufficient to show that $\#(A) \neq \#(P(A))$. Suppose $\#(A) = \#(P(A))$, that is, $A \approx P(A)$. Then there exists a function $g : A \to P(A)$, which is a bijection. Let

$$B = \{x \in A : x \notin g(x)\}.$$

It is clear that $B \subseteq A$, $B \in P(A)$. Since g is onto $P(A)$, there exists $a \in A$ such that $g(a) = B$. So, if $a \in B$, then $a \notin g(a) = B$, meanwhile, if $a \notin B$, then $a \in g(a)$, which implies $a \in B$. Thus, we obtain contradictions in both possible cases.

It means that there does not exist a bijection between A and $P(A)$, equivalently, $\#(A) \neq \#(P(A))$. Therefore,

$$\#(A) < \#(P(A)).$$

It is not difficult to check that if a set A is finite, then

$$\#(A) < 2^{\#(A)} = \#(P(A)).$$

Also, by the above theorem, one can have

$$\#(\mathbb{N}) = \aleph < c = \#(\mathbb{R}),$$

and hence,

$$\#(P(\mathbb{N})) = c = \#(\mathbb{R}),$$

implying that the cardinalities of the power sets of infinitely countable sets are identified with continuum c, and they are understood to be the cardinalities of uncountable sets. Equivalently, that

$$2^{\aleph} = c,$$

in set $\mathscr{C}$ of all cardinal numbers.

Lemma 5.1.53 Let A be a subset of a set B, and let $f : B \to A$ be one-to-one function. Then, for each $X \subseteq (B - A)$, there exists a bijection,

$$f_0 : B \to A \cup X.$$

Proof Let $f_1 : X \to X$ be a function satisfying

$$f_1(X) = X,$$

and let $f_2 : X \to A$ be a function satisfying

$$f_2(X) = f(f_1(X)).$$

Define $f_{i+1} : X \to A$ by functions satisfying the recurrence relation,

$$f_{i+1}(X) = f(f_i(X)), \quad i = 1, 2, \ldots, n.$$

Let

$$C = \bigcup_{i=1}^{\infty} f_i(X).$$

It is clear that $C = f(C) \cup X$. Now, define the function

$$f_0 : B \to A \cup X \text{ by}$$

$$f_0(b) = \begin{cases} b, & \text{if } b \in C, \\ f(b), & \text{if } b \in B - C. \end{cases}$$

The function f_0 is onto, since

$$\begin{aligned} f_0(B) &= f_0(C \cup (B - C)) \\ &= f_0(C) \cup f_0(B - C) \\ &= C \cup f(B - C) \\ &= X \cup F(C) \cup f(B - C) \\ &= X \cup f(B) \\ &= X \cup A. \end{aligned}$$

Also f_0 is one-to-one, because $f_0|_C$ and $f_0|_{B-C}$ are one-to-one functions.

In addition,

$$f_0(C) \cap f_0(B - C) = \phi.$$

Therefore, the function $f_0 : B \to A \cup X$ is a bijection.

Theorem 5.1.54 (Schröder–Bernstein) Let A and B be sets. If A is equipotent to a subset of B, and B is equipotent to a subset of A, then

$$A \approx B.$$

Proof Let $g : A \to B_1$ be a bijection, where $B_1 \subseteq B$, and $h : B \to A_1$ a bijection, where $A_1 \subseteq A$. Then the function

$$(g \circ h) : B \to (g \circ h)(B)$$

is a bijection.

Let $S = B_1 - (g \circ h)(B)$. Then, by the above lemma, there exists a bijection between B and B_1, where

$$B_1 = (g \circ h)(B) \cup S.$$

Hence

$$B \approx B_1,$$

but $B_1 \approx A$. Consequently $A \approx B$.

Corollary 5.1.55 If $\alpha, \beta \in \mathscr{C}$ are cardinal numbers, then

$$\text{if } \alpha \leq \beta \text{ and } \beta \leq \alpha, \text{ then } \alpha = \beta.$$

Proof The proof is done by the very above theorem.

5.2 AXIOM OF CHOICE

In 1954, Zermelo clarifies that there is an axiom that is used implicitly in mathematics. This axiom, called the axiom of choice, is not derived from any previous axioms in mathematics. We here emphasize that there are "many" "equivalent" forms, or statements, of the axiom of choice in various different mathematics fields. The readers can find different expressions of this axiom in many different texts. All of those equivalent forms of "the" axiom of choice say the same meaning:

we can choose certain elements in sets.

Here, we introduce the following form of the axiom of choice.

The Axiom of Choice. If $\{A_i\}_{i\in I}$ is a family of nonempty sets, then there exists a function,

$$f : I \to \bigcup_{i\in I} A_i,$$

such that $f(i) \in A_i, \forall\, i \in I$.

It is clear that, if the index set $I \approx \mathbb{N}_k$ is a finite set for some $k \in \mathbb{N}$, then one can choose $a_1 \in A_1$, $a_2 \in A_2$, ..., $a_k \in A_k$. While, if I is an infinite set, then the axiom will be opaque.

Definition 5.2.1 Let A be a set and let

$$P'(A) = P(A) - \{\phi\}.$$

The function $f : P'(A) \to A$, given by

$$f(B) \in B, \forall\, B \in P'(A),$$

is called a choice function. Sometimes, we write f_B instead of $f(B)$.

Such a choice function f is well-defined by the axiom of choice. That is, one can choose $f(b)$ as an element of B, for $b \in B$, and for all $B \in P(A)$.

Example 5.2.2 Let $A = \{a,b\}$. As an example of the choice function on A, we present a function,

$$f : P'(A) \to A,$$

where

$$f(\{a,b\}) = a,$$
$$f(\{a\}) = a,$$
$$f(\{b\}) = b.$$

Exercises

1. Arrange the following cardinal numbers in order:

$$\#(\{0,1\}), \#([0,1]), \#(\{0\}), \#(P(\mathbb{R})), \#(\mathbb{Q}), \#(\phi)$$
$$\#(P(P(\mathbb{N}_k))), \#(\mathbb{R}), \#(\mathbb{R}-\mathbb{Q}).$$

2. Prove that if $n \in \mathbb{N}$, then $n < \aleph$.

3. Let α, β, and γ be cardinal numbers. Prove that if $\alpha \leq \beta$, then

   (a) $\alpha^\gamma \leq \beta^\gamma$,

   (b) $\gamma^\alpha \leq \gamma^\beta$,

   (c) $\alpha + \gamma \leq \beta + \gamma$.

   (d) $\alpha\gamma \leq \beta\gamma$.

4. Let α, β, and γ be cardinal numbers. Prove that

   (a) $\alpha\beta = 0 \Rightarrow \alpha = 0 \vee \beta = 0$,

   (b) $\alpha\beta = 1 \Rightarrow \alpha = 1 \wedge \beta = 1$.

5. Let α, β, and γ be cardinal numbers. Prove that

$$\alpha \leq \beta \Leftrightarrow \exists \gamma, \text{ such that } \beta = \alpha + \gamma.$$

6. Let α, β, γ, and δ be cardinal numbers, such that $\alpha \leq \gamma, \beta \leq \delta$. Prove that

   (a) $\alpha^\beta \leq \gamma^\delta$,

   (b) $\alpha\beta \leq \gamma\delta$.

7. Let α, β, and γ be cardinal numbers. Prove that

   (a) $\alpha\beta < \alpha\gamma \Rightarrow \beta < \gamma$,

   (b) $\alpha + \beta \leq \alpha + \gamma \Rightarrow \beta < \gamma$.

Bibliography

1. Bittinger, M.L., *Logic, Proof and Sets*, Addison-Wesley, Boston, 1982.

2. Cohen, P.J., *Set Theory and The Continuum Hypothesis*, Dover Publications, Inc., New York, 2008.

3. Dummit, D. S. and Foote, R. M., *Abstract Algebra*, 2nd Edition, Wiley India Pvt. Ltd., New Delhi, 2008.

4. Hall, M., *The Theory of Groups*, Dover Publications, Inc., New York, 2018.

5. Kamke, E., *Theory of Sets*, Dover Publications, Inc., New York, 2003.

6. Senthil Kumar, B. V. and Dutta, H., *Discrete Mathematical Structures: A Succinct Foundation*, 1st Edition, CRC Press, Boca Raton, 2019.

7. Lipschutz, S., *Theory and Problems of Set Theory and Related Topics*, Schaum Publishing Co., New York, 1964.

8. Malik, S.B., *Basic Number Theory*, Vikas Publishing House Pvt. Ltd., New Delhi, 1996.

9. Rosen, K., *Discrete Mathematics and Its Applications with Combinatorics and Graph Theory*, 7th Edition, McGraw Hill Education, New York, 2017.

10. Smith, D., and Andre, R. St., *A Transition to Advanced Mathematics*, 5th Edition, S.Chand (G/L) & Company Ltd, New Delhi, 2001.

11. Spence, L.E., and Eynden, C.V., *Elementary Abstract Algebra*, Harpercollins College Div, New York,1993.

12. Suppes, P., *Axiomatic Set Theory*, Dover Publications, Inc., New York, 2012.

Index

For Product Safety Concerns and Information please contact our EU
representative GPSR@taylorandfrancis.com
Taylor & Francis Verlag GmbH, Kaufingerstraße 24, 80331 München, Germany